Hydroxytriazenes and Triazenes

Hydroxytriazenes and Triazenes

The Versatile Framework, Synthesis, and Medicinal Applications

A.K. Goswami

K.L. Ameta

Shahnawaz Khan

CRC Press

Taylor & Francis Group

Boca Raton London New York

CRC Press is an imprint of the
Taylor & Francis Group, an **informa** business

First edition published 2021
by CRC Press
6000 Broken Sound Parkway NW, Suite 300, Boca Raton, FL 33487-2742

and by CRC Press
2 Park Square, Milton Park, Abingdon, Oxon, OX14 4RN

© 2021 Taylor & Francis Group, LLC

CRC Press is an imprint of Taylor & Francis Group, LLC

ISBN: 978-1-138-59720-4 (hbk)
ISBN: 978-0-367-55594-8 (pbk)
ISBN: 978-0-429-48705-7 (ebk)

Typeset in Palatino
by Deanta Global Publishing Services, Chennai, India

*Dedicated to Professor D.N. Purohit, who taught
me the alpha to omega of hydroxytriazenes.*

A.K. Goswami

Contents

Preface

Nitrogen-based compounds are a most important member of organic chemical compounds. A significant fraction of current research in this area has now moved toward the N-heterocyclic compounds containing nitrogen because a large number of compounds, both synthetic and natural, bearing nitrogen-containing simple (pyrrole, pyrimidine, indole, quinoline, and purine) rings and fused heterocyclic skeletons (4-anilinoquinazolines, pyrazolopyrimidines, triazolopyrimidines, pyrrolopyrimidines, pyrazolopyridazines, and imidazopyrazines), have been discovered with excellent bioactivity, many of which are presently in clinical trials [1–5]. The novel, cost-effective, and sustainable synthetic methods for the synthesis of N-heterocyclic compounds have always been among the most significant research areas in synthetic chemistry. In the past decades, numerous conventional methods for the preparation of N-heterocycles have been improved to meet the enhanced demands of modern organic synthesis and medicinal chemistry [6]. The triazine scaffold is prevalent in a variety of pharmacologically active synthetic and natural compounds. On the other hand, triazine compounds, like dacarbazine and temozolomide, are a group of alkylating agents with similar chemical, physical, antitumor, and mutagenic properties and are of wide medicinal interest. Dacarbazine requires hepatic activation whereas temozolomide is spontaneously converted into an active metabolite in aqueous solution at physiological pH. Moreover, temozolomide is fully active when administrated orally (100% bioavailability). Further, triazene compounds have excellent medicinal properties and limited toxicity. Hydroxytriazenes [7] are an important class of compounds having R–N–N=N–R' framework with an –OH group at 3 –position. They are excellent chelating agents for transition metals and have been exhaustively studied for both the complexometric and spectrophotometric estimation of transition metals and some non-transition metals. In recent years, they have also been studied from a biological viewpoint and have shown very promising biological and medicinal applications. Although sporadic publication on their synthetic as well as medicinal applications is reported, no consolidated information on this class of compounds is available for researchers. It is worth highlighting their use as excellent and stable synthetic intermediate moieties transformable to interesting biological precursors, as well as the possibility of exploring their medicinal application for activities as diverse as antibacterial, antifungal, antioxidant, anti-inflammatory, antidiabetic, and even cytotoxic activities to name a few. In addition, their simple and green synthesis with a small framework and the use of mostly aqueous media, and minimum or no non-aqueous solvents make them interesting compounds for both synthetic organic and medicinal chemists. On the other hand, triazenes, $R^1R^2N-N=NR^3$, are one of the simplest frameworks of the azene class of hydronitrogen compounds not found in nature. The class holds much promise in synthetic chemistry as it consists of reactive groups which are both stable and adaptable to numerous synthetic transformations. The compounds have been known for over 100 years and are versatile as they have found different applications as chemical and biological reagents, pharmacological, total synthesis, polymer technology, and even for the construction of novel ring systems. This book highlights recent advances and diverse possible applications such as biological and environmental functions. Thus, the book includes holistic information on synthetic methods for novel compounds based on this moiety, up to date information on the how and why of their diverse or even multitargeted medicinal applications, and future state of the art of both aspects. The text is a valuable resource, which will provide guidance for future researchers of medicinal chemistry.

References

1. Kidwai, M., Venkataramanan, R., Mohan. R., and Sapra, P. Cancer chemotherapy and heterocyclic compounds. *Curr. Med. Chem.* 2002, 9, 1209.
2. Cozzi, P., Mongelli, N., and Suarato, A. Recent anticancer cytotoxic agents. *Curr. Med. Chem.—Anti-Cancer Agents* 2004, 4, 93.
3. Prudhomme, M. Biological targets of antitumor indolocarbazoles bearing a sugar Moiety. *Curr. Med. Chem.—Anti-Cancer Agents* 2004, 4, 509.
4. Scozzafava, A., Owa, T., Mastrolorenzo, A., and Supuran, C. T. Anticancer and antiviral sulfonamides. *Curr. Med. Chem.* 2003, 10, 925.
5. Belmont, P., Bosson, J., Godet, T., and Tiano, M. Acridine and acridone derivatives, anticancer properties and synthetic methods: Where are we now? *Anti-Cancer Agents Med. Chem.* 2007, 7, 139.
6. Xu, W., and Fu, H. Amino acids as the nitrogen-containing motifs in copper-catalyzed domino synthesis of N-heterocycles. *J. Org. Chem.* 2011, 76, 3846.
7. (a) Dalawat, D. S., Chauhan, R. S., and Goswami, A. K. Review of spectrophotometric methods for determination of zirconium. *Rev. Anal. Chem.* 2009, 24(2), 75–102. (b) Khanam, R., Singh, R., Mehta, A., Dashora, R., and Chauhan, R. A review of spectrophotometric methods for determination of nickel. *Rev. Anal. Chem.* 2005, 24(3), 149–245. (c) Singh, K., Chauhan, R. S., and Goswami, A. K. A review of reagents for spectrophotometric determination of lead (II). *Main Group Metal Chem.* 2005, 28(3), 119–148. (d) Upadhyay,

M., Chauhan, R. S., and Goswami, A. K. A review of spectrophotometric reagents for cadmium determination. *Main Group Metal Chem.* 2005, 28(6), 301–357. (e) Khan, S., Dashora, R., Goswami, A. K., and Purohit, D. N. A review of spectrophotometric methods for determination of iron. *Rev. Anal. Chem.* 2004, 23(1), 1–74. (f) Kumar, S., Goswami, A. K., and Purohit, D. N. A review of hydroxytriazenes. *Rev. Anal. Chem.* 2003, 22(1), 73–80.

Authors

A.K. Goswami, PhD (Retired Professor), is a UGC BSR Fellow who worked in the Department of Chemistry, M.L. Sukhadia University, Udaipur, India, for more than 33 years in various capacities. He is actively working on the medicinal applications of hydroxytriazenes and their metal complexes. He has contributed enormously in this field having published more than 130 research papers and 10 reviews in addition to several books. He has supervised 30 doctoral students for their PhD. He served in the Department of Chemistry, Kenyatta University, Nairobi, for two years as visiting faculty and has presented his research in countries such as the United States, Japan, South Africa, and Kenya, to mention a few. He is Vice President of the Indian Society for Chemists and Biologists and is President of the Society for Promotion of Science, Education, and Research (SPSER) where he is presently working on projects related to teaching and popularizing science in rural areas and economically weaker sections of society.

K.L. Ameta, PhD, is a Professor in the Department of Chemistry, School of Liberal Arts and Sciences, Mody University of Science and Technology, Lakshmangarh, Rajasthan, India. He is also the Coordinator for Research and Development, SLAS at Mody University. He received his doctorate degree in organic chemistry from M.L. Sukhadia University, Udaipur, India, in 2002. He has vast experience in teaching both undergraduate- and postgraduate-level students. His research area involves the synthesis, characterization, and biological evaluation of different-sized bioactive heterocyclic systems. In addition, he has keen interests in heterogeneous catalyzed organic synthesis and photocatalysis. He has published about 60 research articles in journals of national and international repute.

Shahnawaz Khan, PhD, is an Assistant Professor in the Department of Chemistry, Prince College, Sikar. He completed his PhD from the Medicinal and Process Chemistry Division, CSIR–Central Drug Research Institute, Lucknow, India (a National Federal Research Laboratory of the Government of India). Dr. Khan worked as a Prestigious D.S. Kothari Postdoctoral Fellow in the Department of Chemistry, University of Rajasthan, during 2014–2017. He has published more than 30 research papers in international journals of repute. Dr. Khan has participated in various conferences of national and international repute.

chapter one

Hydroxytriazenes
State of the art and future prospects

Metal determination and detection have been challenges ever since the development of analytical chemistry. There is no need to explain the usefulness of metals both in bulk and traces to humankind. Not only that, but both medicinal applications and the use of metals and metal complexes are emerging fields of chemistry as well as the pharmaceutical sciences. Several monographs and reviews, dedicated volumes and publications evidence the significance of the discipline [1–5]. Hydroxytriazenes are a group of chelating agents well-known for their analytical application in the detection and determination of entire d-series in general and first transition series in particular. Their application as spectrophotometric and complexometric reagents is fully established; voluminous reviews on their spectrophotometric and other applications have appeared during the last couple of years [6–11].

1.1   Introduction

Shome [12] made the first attempt to remove defects of cupferron (ammonium salt of nitrophenyl hydroxylamine), a reagent widely used then for the estimation and separation of different metal ions. Since the reagent itself and the complexes prepared using cupferron were thermally unstable, photosensitive, and even had poor shelf life for storage, the attempt made by Shome led to an interesting series of new reagents. By replacing the nitro group with a benzoyl group or even by other substituents he examined the new reagent for better application devoid of the drawbacks of the earlier reagent. Nonetheless, Bamberger [13] and Gebhard and Thompson [14] synthesized a number of hydroxytriazenes for the first time. Subsequently, the excellent complexing ability of these compounds was examined by Elkins and Hunter [15].

The names diazohydroxyamine compounds and triazene oxides have been given in *Chemistry of Carbon Compounds* edited by Rodd [16]. However, Sogani et al. preferred to use 3-hydroxytriazene as the name of this family of compounds. The common functional grouping of this class of compounds is represented in Figure 1.1.

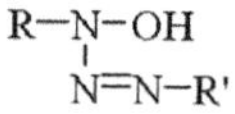

$$R-N-OH$$
$$N=N-R'$$

Figure 1.1 Where R can be an alkyl/aryl group and R' an phenyl group.

A number of excellent reviews has been published by Purohit [17] and Chakraborty and Majumdar [18] on the analytical applications of this class of organic reagents. It is evident from the large number of publications and reviews appearing on this class of compounds that they have been exhaustively used for the spectrophotometric methods of metal estimation, including of transition metals and some non-transition metals also. In this development, two voluminous reviews by Purohit et al. and Deepika K. Gorji et al. [20] have been exclusively written on hydroxytriazenes and their different uses in addition to uses as spectrophotometric reagents for metal estimation. Further, reviews on spectrophotometric reagents for various metals such as iron, copper, chromium, cadmium, zirconium, nickel, cobalt, palladium, zinc, vanadium, etc., also incorporates hydroxytriazenes [21–29]. Even a review on lead includes hydroxytriazenes as reagents. The volume of reviews and publications itself warrants a systematic description of this interesting series of compounds.

The application of hydroxytriazenes apart from their use as spectrophotometric reagents or even metallochromic indicators in complexometric determination dates back to the early 1960s. A number of U.S. patents are available in the literature which establish their application as bioactive molecules [30]. A patent by Gubler [31] reported the use of hydroxytriazenes for controlling insects and arachnids. However, the area of their medicinal application is still an underexplored one.

1.2 State of the art of hydroxytriazenes— Spectrophotometric reagent and medicinal/bioactive molecules

A series of hydroxytriazenes with varied substituents has been offering excellent spectrophotometric as well as complexometric reagents. Almost an entire transition series can be successfully estimated using these reagents. More than 300 compounds have been synthesized and used for their application as analytical reagents. A review of hydroxytriazenes used for this application would establish this fact. An entire discussion on this is included in a subsequent section of this chapter. This section incorporates the most recent developments particularly in the last ten years. A number of voluminous reviews on hydroxytriazenes and other spectrophotometric reagents for specific metals have already appeared and been published by our group. The medicinal application of hydroxytriazenes on the other hand is of recent origin. Although sporadic publications and patents appear in the literature, no systematic study on these is available. The section subsequent to that on spectrophotometric application will address this area and will include the present scenario at an introductory level. The beauty of these compounds is that they offer a variety of medicinal applications by the variation in their structures and they are very simple to synthesize. Their activities include antibacterial, antifungal, insecticidal, acaricidal, anti-inflammatory, antioxidant, anti-diabetic, and even antitumor activities. The detailed survey of the published work in recent years will highlight this fact.

1.3 A review of hydroxytriazenes as analytical reagents

Gorji et al. [20] have reviewed the applications of hydroxytriazenes over the years 1990–1997. The review is comprised of work published during these years for different hydroxytriazenes and transition metals. A number of hydroxytriazenes have been used for the estimation of transition metals since the year 1999, and no consolidated review on the work is available. To save space and to avoid going beyond the scope of this book, a table of hydroxytriazenes which has been used recently for the estimation of individual transition metal is supplied. Table 1.1 includes the structure of each hydroxytriazene and brief details of the method, along with the reference of the work. This will help researchers working in the area of application. The simplest approach to plan synthesis based on molecular structure can be done in silico and the predicted activity verified experimentally, by using some of the available software. It saves a lot of costly synthesis, testing and failure of the molecule in terms of activity. Although this scheme is at a very basic level since its success rate can also be not high enough, it can prove to be a very good strategy in developing new drugs.

1.4 Medicinal applications of hydroxytriazenes: An underexplored area

A survey of the literature reveals that some patents on medicinal applications of hydroxytriazenes are available but no work has been done exclusively on the development of these molecules as drugs. There could have been issues such as toxicity, a lack of detailed study on structure–activity relationship (SAR), or even a randomness of approach of application of analytical reagents to this non-conventional use as potential drug-precursors or drugs.

Although computer-aided drug design is of recent origin and the success rate based on structure–activity relationship is not high, it has helped a lot in the development of novel molecules not reported earlier for their drug candidature.

Table 1.1 Review of hydroxytriazenes as analytical reagents

Metal	Name of the reagent	λ_{max} of working wavelength/pH	Remarks/method	Ref
Co(II) Cu(II)	3-hydroxy-3-(4-mephenyl)-1-(3-chloro, 2-me phenyl) triazene	pH 6.7–7.2	Simultaneous determination of Cu(II) and Co(II) ions by H-point standard method	[32]
Ni(II)	3-hydroxy-3-phenyl-1-dichloro phenyl) triazene	6.5–7.0	Polarographic method	[33]
	3-hydroxy-3-n-propyl-1-1-(4-sulfonamidophenyl) triazene	400 nm/6.9–7.3	Spectrophotometry	[34]
	3-hydroxy-3-isopropyl-1-(4-sulfonamidephenyl) triazene	395/6.7–7.3	Spectrophotometry	[35]
Cu(II)	3-hydroxy-3-phenyl-1-o-nitrophenyltriazene	5.5–6.5	Complexometry	[36]
Mo(VI)	3-hydroxy-3-p-Tolyl-1-p-carboxyphenyltriazene	2.0–4.0	Spectrophotometry	[37]
Co(II)	3-hydroxy-3-m-tolyl-1-p-sulfonato (sodium salt) phenyltriazene	7.5–9.5	Polarography	[38]
Cu(II)	3-hydroxy-3-m-tolyl-1-p-(sulfonamido) phenyltriazene	6.5–7.1	Polarography	[39]
Cu(II)	1,3-diphenyl-3-hydroxytriazene		Polarography	[40]
Ni(II)	1-(6-hydroxy-2-pyrine)-3-(4-phenylazo-phenyl)-triazene	549 nm/10.97	Spectrophotometry	[41]
Ni(II)	5-hydroxy-2-sulfo-4-nitrophenyl-diazoaminoazobenzene	528 nm/10.0	Spectrophotometry	[42]
Cu(II)	3-hydroxy-3-methyl-1-p-sulfonamidophenyl triazene		Complexometry	[43]
Ni(II)	3-hydroxy-3-phenyl-1-o-trifluorophenyltriazene	410 nm/8.5–	Spectrophotometry	[44]
V(V)	3-hydroxy-3-p-chlorophenyl-1-p-nitrophenyltriazene	410 nm/7.0–9.0	Spectrophotometry	[45]
Cd(II)	2-hydroxy-4-sulfamybenzene-3-(4-nitrophenyl)triazene	535 nm/11.0–12.0	Spectrophotometry	[46]
Cd(II)	1-(6-hydroxy-2-pyrine)-3-(4-phenylazophenyl)triazene	530 nm/9.0–11.0	Spectrophotometry	[47]
Co(II)	3-hydroxy-3-methyl-1-(4-sulfonamidophenyl) triazene	360 nm/6.7–7.3	Spectrophotometry	[48]
Cd(II)	2-hydroxy-5-nitrophenyldiazoaminoazobenzene	8.5–10.0	Spectrophotometry	[49]
Zn(II)	1-(2-hydroxy-3,5-dinitropheynyl)-3-(4-phenylazophenyl)-triazene	525 nm/11–12	Spectrophotometry	[50]
Cd(II) Pb(II) Zn(II) Co(II)	3-hydroxy-3-phenyl-1-p-carboxyphenyltriazne		Polarography	[51]

1.5   Drug likeliness—A theoretical concept in drug design

The drug likeliness is a quantitative concept for designing a drug which is based on a number of factors and basically bioavailability. The basis of drug likeliness is the molecular structure of the substance which can be tested well before the synthesis.

We will describe the approach which has been applied to hydroxytriazenes by our group. This approach makes use of theoretically predicted structures and possible use of a particular bioactivity. Prediction of activity spectra of substances (PASS) is a very simple tool available for this. Details of studies based on theoretical prediction for hydroxytriazenes will be incorporated in a subsequent chapter.

Modern drug discovery and development is central part of medicinal chemistry. During the 20th century, a majority of drugs wore discovered. The roots of medicinal chemistry stretch to a number of branches of chemistry. Recent developments in medicinal chemistry have focused on synthesizing new compounds and understanding the molecular basis of disease. The biomolecular targets which cause disease are identified, and how the specific compounds or hits can block the identified disease-causing biomolecule is the modern scheme of development of drugs. SAR is a very useful tool to improve the efficacy of hits. Hydroxytriazenes have been successfully developed as medicinally applicable compounds by using the strategy of theoretical prediction of a new structure with a hydroxytriazene moiety, synthesizing these compounds and validating some of the predicted activities. Thus, the next chapter will include our recent research highlighting this scheme and the medicinal application of the compounds.

References

1. Fawell, N. P. *Transition Metal Complexes as Drugs and Chemotherapeutic Agents*. James, B. R., and Ugo, R., eds. Reidel Kluwer Academic Press, Dordrecht, 1989, 11.
2. Farrel, N. P. *The Uses of Inorganic Chemistry in Medicine*, Royal Society of Chemistry, Cambridge, 1999.
3. Geo, Z., and Sadler, P. J. Metals in medicine. *Angew Chem. Int. Ed. Engl.* 1999, 38, 1512–1531.
4. Keppler, B. *Metal Complexes in Cancer Chemotherapy*. VCH, Basel, 1993.
5. Fricker, S. P. *Metal Complexes in Cancer Therapy*. Chapman & Hall, London, 1994, 1, 215.
6. Purohit, D. N. Hydroxytriazenes: A review of a new class of chelating agents. *Talanta* 1967, 14, 353.
7. Chakravorty, D., and Majumdar, A. K. Hydroxytriazenes as chelating agents in analytical chemistry. *J. Indian Chem. Soc.* 1977, 54, 238.

8. Dutta, R. L., and Sharma, R. Coordination-complexes of triazene-1-oxides. *J. Sc. Ind. Res. India* 1981, 40, 715.
9. Purohit, D. N., Nizamuddin, and Golwalkar, A. M. Hydroxytrazenes as chelating agents in analytical chemistry. *Revs. Anal. Chem. (Israel)* 1985, 8, 76.
10. Purohit, D. N., Tyagi, M. P., Bhatnagar, R., and Bishnoi, I. R. Hydroxytriazenes as chelating agents: A review. *Revs. Anal. Chem. (Israel)* 1992, 11, 269.
11. Goswami, A. K., Chauhan, R. S., Dashora, Rekha, Mehta, Anita, Singh, Ravindra, and Khanam, Rehana. Review of spectrophotometric methods for determination of nickel. *Revs. Anal. Chem. (Israel)* 2005, 24(3), 149.
12. Shome, S. C. Gravimetric determination of copper, iron, aluminium and titanium with N-benzoylphenylhydroxylamine. *Analyst* 1950, 75, 27.
13. (a) Bamberger, E. Ueber die Einwirkung des nitrosobezole auf Amidoverbin. *Ber.* 1896, 26, 104.
    (b) Bamberger, E., and Renauld, E. Ueber alphyllirte und alkylirte hydroxylamine. *Ber.* 1897, 30, 2285.
    (c) Bamberger, E., and Busdorf, W. Ueber die Einwirkung von Nitrosbenzol auf aromatische hydrazine. *Ber.* 1900, 33, 3510.
14. Gebhard, N. L., and Thompson, H. B. Diazohydroxylamino compounds and the influence of substituting groups on the stability of their molecules. *J. Chem. Soc. (London)*, 1907, 767.
15. Elkins, M., and Hunter, L. The azogroup as a chelating group, Part IV The constitution of the Arylazo-bisoximes. *J. Chem. Soc. (London)* 1938, 1346.
16. E. H. Rodd, ed., *Chemistry of Carbon Compounds*, Elsevier, New York, 1951, Vol. I, pp. 532.
17. Purohit, D. N. Hydroxytriazenes: A review of a new class of chelating agents. *Talanta* 1967, 14, 353.
18. Chakraborty, D., and Majumdar A. K. Hydroxytriazenes as chelating agents in analytical chemistry. *J. Indian Chem. Soc.* 1977, 54, 258.
19. Purohit, D. N., Nizamuddin, Golwalkar, and Arun M. Hydroxytrazenes as chelating agents in analytical chemistry. *Revs. Anal. Chem. (Isreal)* 1985, 8, 76.
20. Gorji, Deepika K., Chauhan, R. S., Goswami, A. K., and Purohit, D. N. Hydroxytriazenes: A review. *Revs. Anal Chem. (Israel)* 1998, 16(4), 223.
21. Kumar, S., Goswami, A. K., and Purohit, D. N. A review of Hydroxytriazenes. *Rev. Anal. Chem.* 2003, 22(1), 73.
22. Rezare, Behrooz, Zabeen, Raziya, Goswami, A. K., and Purohit, D. N. Review of spectrophotometric methods for determination of vanadium. *Revs. Anal. Chem.* 1993, XII (1–2), 1.
23. Ressalan, S., Chauhan, R. S., Goswami, A. K., and Purohit, D. N. Review of spectrophotometric methods for determination of chromium. *Rev. Anal. Chem.* 1997, 16(2), 69.
24. Ram, G., Chauhan, R. S., Goswami, A. K., and Purohit, D. N. Review of spectrophotometric methods for determination of cobalt(II). *Rev. Anal. Chem.* 2003, 22(4), 255.
25. Khan, S., Dashora, R., Goswami, A. K., and Purohit, D. N. Review of spectrophotometric methods for determination of iron. *Rev. Anal. Chem.* 2004, 23(1), 1.

26. Upadhyay, M., Chauhan, R. S., and Goswami, A. K. A review of spectrophotometric reagents for cadmium determination. *Main Group Metal Chem.* 2005, 28(6), 301.

27. Singh, K., Chauhan, R. S., and Goswami, A. K. A review of reagents for spectrophotometric determination of lead (ii). *Main Group Metal Chem.* 2005, 28(3), 119.

28. Khanam, R., Singh, R., Mehta, A, Dashora, R., Chauhan, R. S., and Goswami, A. K. Review of spectrophotometric methods for determination of nickel. *Rev. Anal. Chem.* 2005, 24(3), 149.

29. Dalawat, D. S., Chauhan, R. S., and Goswami, A. K. Review of spectrophotometric methods for determination of zirconium. *Rev. Anal. Chem.* 2005, 24(2), 75.

30. Forster, Heinz. Triazene substituted diphenyl derivative are suitable as chemotherapeutic agents for treating carcinomas in humans and animals. Patent U.S. 9409859B2. https://patents.google.com

31. Gubler, K. *Certain 3-Hydroxytriazene and Their Use of Controlling Insects and Arachnids.* Patent U.S. 3714351A. https://patents.google.com

32. Panwar, A. K., Choubisa, N. K., Kodli, K. K., Regar, M., Dashora, R., Chauhan, R. S., and Goswami, A. K. Simultaneous determination of copper (II) and cobalt (II) ions by H-point standard addition method. *IOSR J. Appl. Chem.* 2015, 8(2–2), 22–28.

33. Kodli, Krishan Kant, Choubisa, Naresh, Panwar, Ashok, Dashora, Rekha, Chauhan, R. S., and Goswami, A. K. Polarographic studies of 3-hydroxy-3-phenyl-1-(dichloro phenyl) triazene with Ni (II) complex. *Rasayan J. Chem.* 2015, 8(2), 172–175.

34. Khanam, Rehana, Khan, Saba, and Dashora, Rekha. 3-Hydroxy-3-N-propyl-1-(4-sulphonamidophenyl) triazene: A new reagent for spectrophotometric determination of nickel (II). *Orient J. Chem.* 2014, 30(2), 837–841.

35. Khanam, Rehana, Khan, Saba, Dashora, Rekha, Chauhan, R. S., and Goswami, A. K. Analytical application of 3-hydroxy-3-isopropyl-1-(4-sulphonamidophenyl) triazene in the spectrophotometric determination of nickel (ii). *Int. J. Pharm. Chem. Biol. Sci.* 2013, 3(3), 704–707.

36. Singh, R. P., Bhandari, Amit, Mehta, Anita, Chauhan, R. S., and Goswami, A. K. Polarographic study of cobalt (II) complex with 3-hydroxy-3-o-toly-1-psulphonato (sodium salt) phenyltriazene. *Int. J. Chem. (Mumbai, India)* 2012, 1(2), 194–198.

37. Babel, Tushita, Bhandari, Amit, Jain, Preksha, Mehta, Anita, and Goswami, A. K. Synthesis , activity prediction and spectrophotometric study of molybdenum complex of 3-hydroxy-3p-tolyl-1p-carboxyphenyltriazene. *Int. Res. J. Pharm.* 2012, 3(5), 382–385.

38. Singh, Girdhar Pal, Joshi, Pooja, Pareek, Neelam, Upadhyay, Dipen, Bhandari, Amit, Chauhan, R. S., and Goswami, A. K. Polarographic determination of Co (II) complex with 3-hydroxy-3-m-tolyl-1-p-sulphonato(sodium salt) phenyltriazene. *Rasayan J. Chem.* 2010, 3(4), 690–92.

39. Joshi, P., Pareek, N., Upadhyay, D., Khanam, R., Bhandari, A., Chauhan, R. S., and Goswami, A. K. Polarographic determination and antifungal activity of Cu(II) complex with 3-hydroxy-3-m-tolyl-1-p-(sulphonamido) phenyltriazene. *Int. J. Pharm. Sci. Drug Res.* 2010, 2(4), 278–280.

40. Gupta, Manu, Bairwa, B. S., Kamawat, Romila, Sharma, I. K., and Verma, P. S. Electrochemical behaviour of 1, 3-diphenyl-3-hydroxytriazene and its copper complex. *Indian J. Chem.* 2008, 47A(3), 383–386.

41. Zhang, Yan-Hong, Fan, Yue-Qin, Guo, Yong, Feng, Feng, Liu, Yong-Wen, Meng, Shuang-Ming, and Liu, Hul-Jun. Spectrophotometric determination of micro nickel with 1-(6-hydroxy-2-pyrine)-3-(4-phenylazo-phenyl)-triazene. *Guangpu Shiyanshi* 2007, 24(4), 649–651.

42. Xu, Lin, Fan, Yuegin, Meng, Shuangming, Guo, Yong, Liu, Yongwen, and Liu, Hongyan. Spectrophotometric determination of Arsenic(v) in surface water. *Yeijin Fenxi* 2006, 28(3), 18–20.

43. Sharma, J. C., Upadhyay, Mishika, Jain, C. P., and Dashora, Rekha. Complexometric determination of copper in pharmaceutical samples using hydroxytriazenes. *Asian J Chem.* 2006, 18(4), 3144–3146.

44. Ghaisvand, Ali Reza, Rezaei, Behrooz, and Masroor, Gholam Abbas. Synthesis of a new hydroxytriazene derivative and its application for selective extraction-spectrophotometric determination of nickel (ii). *Asian J. Chem.* 2006, 18(3), 2185–2193.

45. Rezaei, B., Ghiasvand, A. R., Purohit, D. N., and Rostami, S. Direct spectrophotometric determination of vanadium(v) using a recently synthesized hydroxytriazene derivative. *Asian J. Chem.* 2005, 18(1), 689–690.

46. Zheng, Yun-fa, Zhang, Chun-niu, and Gu, Yong-bing. Applications of Schiff bases and their metal complexes as catalysts. *Zheijang Shifan Daxue Xuebao, Ziran Kexueban* 2005, 28(2), 183–186.

47. Fan, Yue-qin, Meng, Shuang-ming, Gup, Yong, Liu, Yong-wen, Fang, Guo-zhen, Feng, Feng, and Liu, Jian-hong. Synthesis of 1-(6-hydroxy-2-pyrine)-3-(4-phenylazophenyl)triazene and its chromogenic reaction with cadmium. *Fenxi Shiyanshi* 2005, 24(1), 11–13.

48. Naulakha, Neelam, Goswami, A. K. Synthesis and analytical application of 3-hydroxy-3-methyl-1-(4-sulphonamidophenyl)triazene in the spectrophotometric determination of copper(II). *J. Indian Chem. Soc.* 2004, 81(5), 438–439.

49. Zheng, Yunfa, Yang, Minghua, and Shen, Sumel. Effects of β-cyclodextrin and cetyl pyridine bromide on the colour reaction of zinc (II) with α,β,γ,δ-tetrakis(4-trimethylammoniumphenyl) porphyrin. *Huaxue Shiji* 2003, 25(1), 31–32, 49.

50. Feng, Yonglan, and Kuang, Daizhi. Lihua, Jianyan. Color reaction of 1-(2-hydroxy-3,5-dinitrophenyl)-3-(4-phenylazophenyl)-triazene with zinc and its application. *Huasue Fence* 2004, 38(5), 229–230, 233.

51. Pratihar, P. L., Lohiya, R. K., Mukherjee, S. K., and Singh, R. V. Determination of kinetic parameters and polarographic studies of cadmium(ii), lead(ii), zinc(ii) and copper(ii) with 3-hydroxy-3-phenyl-1-p-carboxyphenyl triazine (HPCPT) ion. *Orient. J. Chem.* 2002, 18(1), 15–20.

chapter two

Synthesis, characterization, and structure of hydroxytriazenes

2.1 Synthesis of hydroxytriazenes

A survey of the literature reveals that there are three general methods for preparing hydroxytriazenes:

I. Reaction of phenylhydrazine or substituted phenylhydrazine with aliphatic or aromatic nitro or nitroso compounds [1]
II. Coupling reaction of deazocompounds with alkyl or aryl substituted hydroxylamines [2]
III. By oxidation of diazoaminobenzene with peroxybenzoic acid in a suitable solvent [3]

2.1.1 Method I

This method involves the reaction of nitrobenzene or substituted nitrobenzene with phenyl or substituted phenylhydrazine to obtain hydroxytriazene. The general scheme can be represented by the following reaction.

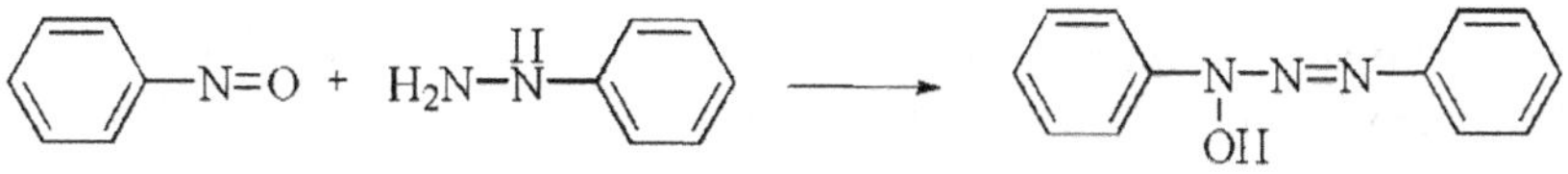

Scheme 2.1

2.1.2 Method II

This method is most convenient and hence a commonly used method for preparing hydroxytriazenes. Ease of preparation and good yield with minimum synthetic assembly are the advantages of this method over the other two methods. The method involves the coupling of an alkyl or aryl hydroxylamine with an aryl diazonium salt at 0–5°C in a 1:1 molar ratio at a neutral pH. The scheme can be represented as follows.

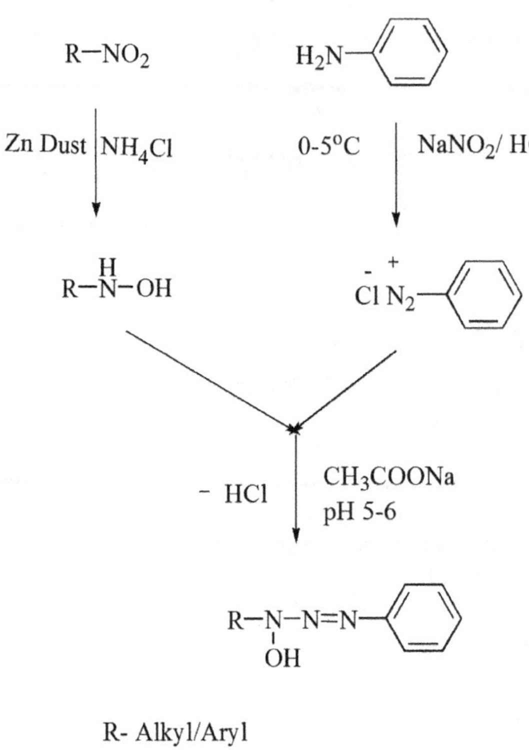

$$R-NO_2 \qquad\qquad H_2N-\!\!\!\bigcirc$$

Zn Dust | NH$_4$Cl $\qquad\qquad$ 0-5°C | NaNO$_2$/ HCl

$$R-\overset{H}{N}-OH \qquad\qquad \overset{-\ +}{Cl\,N_2}-\!\!\!\bigcirc$$

− HCl $\qquad$ CH$_3$COONa
pH 5-6

$$R-N-N=N-\!\!\!\bigcirc$$
$$\underset{OH}{|}$$

R- Alkyl/Aryl

Scheme 2.2

2.1.3   *Method III*

In the third method, the oxidation of diazomino benzene is facilitated with peroxybenzoic acid under mild conditions with a suitable solvent to yield a hydroxytriazene. A general reaction may be represented as follows.

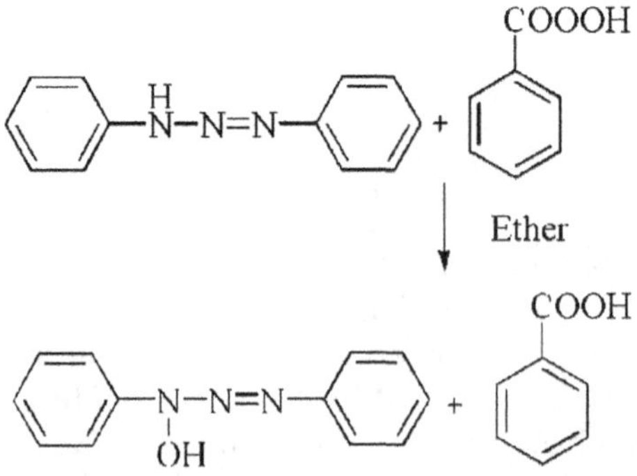

$$\bigcirc\!\!-\!\!\overset{H}{N}-N=N-\!\!\!\bigcirc \;+\; \bigcirc\!\!\overset{COOOH}{|}$$

Ether

$$\bigcirc\!\!-\!\!\underset{OH}{\overset{|}{N}}-N=N-\!\!\!\bigcirc \;+\; \bigcirc\!\!\overset{COOH}{|}$$

Scheme 2.3

It is important to mention here that out of these three synthetic routes, the second one is the most common and advantageous preparative route to obtain hydroxytriazene. The literature evidences the synthesis of a large number of compounds of this series using the second method.

2.2   Characterization of hydroxytriazenes

2.2.1   Spot test detection of hydroxytriazenes

Four spot test detection methods are used to confirm the synthesis of hydroxytriazenes:

  (i) α-naphthylamine test [4]
 (ii) N,N-dimethylaniline test [5]
(iii) Sulfuric acid test [6]
 (iv) Picric acid test [7]

  (i) **α-Naphthylamine test**: To a small amount of hydroxytriazene dis-
      solved in acetic acid (0.001% w/v), α-napthyl amine (0.01% w/v)
      dissolved in acetic acid is added. The reaction immediately gives
      a reddish brown or pink color which confirms the presence of a
      hydroxytriazene group. A gentle warming of this solution is recom-
      mended since this intensifies the color. The development of this color
      is tentatively due to the formation of a tetrazene or a cyclictetrazene.
      The reaction may be represented as follows.

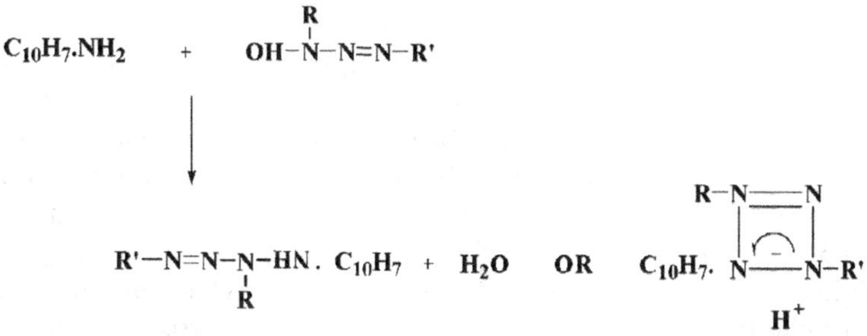

$$C_{10}H_7.NH_2 \quad + \quad \overset{\overset{\displaystyle R}{|}}{OH-N}-N=N-R'$$

$$R'-N=N-\underset{\underset{\displaystyle R}{|}}{N}-HN \cdot C_{10}H_7 \;+\; H_2O \quad OR \quad C_{10}H_7 \cdot \; \underset{\underset{\displaystyle H^+}{}}{\overset{\displaystyle R-N\!=\!\!=\!\!N}{N\!-\!\!-\!\!N-R'}}$$

Scheme 2.4

 (ii) **N,N-Dimethylaniline test**: When a pinch of solid hydroxytriazene,
      few drops concentrated HCl and dimethylaniline are heated to boil-
      ing, a red or reddish violet to deep violet color develops, confirming

the presence of a hydroxytriazene group. A similar test is reported by Coasts and Katritzky [8] for pyridine-1-oxide, which is a general test for N-oxides. Since hydroxytriazenes are also represented as N-oxides the test confirms the presence.

(iii) **Sulfuric acid test**: A very simple test using sulfuric acid is described for detecting hydroxytriazenes. When 1.0 ml of concentrated H_2SO_4 is slowly heated with a pinch of hydroxytriazene a green or reddish brown color appears, confirming the presence of hydroxytriazenes.

(iv) **Picric acid test**: When a pinch of solid hydroxytriazene or a drop of ethanolic solution of hydroxytriazene (0.001% w/v) is treated with two to five drops of a saturated ethanolic solution of picric acid and heated for 1 min over a water bath at around 55–60°C, the development of a red color confirms the presence of hydroxytriazene.

Although color also appears at cold temperatures, heating is recommended. As far as the chemical reaction of the test is concerned both amineoxide ($\equiv N \longrightarrow 0$) and secondary amino (= N–H) groups form picrates which is the reason for the development of a red color. It is due to the formation of picrates which are the donor acceptor type of molecular complexes.

2.3 Structure of hydroxytriazenes and tautomerism

(i) 3-Hydroxy-1,3-diphenyltriazene (HDPT) as the parent compound of the hydroxytriazene class.

The compound prepared by the coupling of phenylhydroxylamine with benzene diazonium chloride is a light yellow solid with m.p. 119–120. It is very stable compound which can be preserved for quite a long time. Most of the complexes prepared with the compound are granular with a definite composition suitable for direct weighing as they are thermally stable. The spectral characteristics of the reagent using alcohol are λ_{max} 238 nm with log $\in$= 4.328. The general structure of hydroxytriazenes or their analogues can be represented by the structure of 3-hydroxy-1,3-diphenyltriazene which has been well-established by Purohit et al. [9,10]. The author has reported on the basis of UV spectral studies of HDPT by Calzolari and Furlani [11] that it exists either in a single form or in two forms which have identical spectra. Purohit et al. [12] have suggested the existence of a hydrogen-bonded form whether it is present in hydroxytriazene or in its tautomeric triazene oxide form. The same conclusion was published by Purohit and Sogani [13] on the basis of the low dissociation constants of

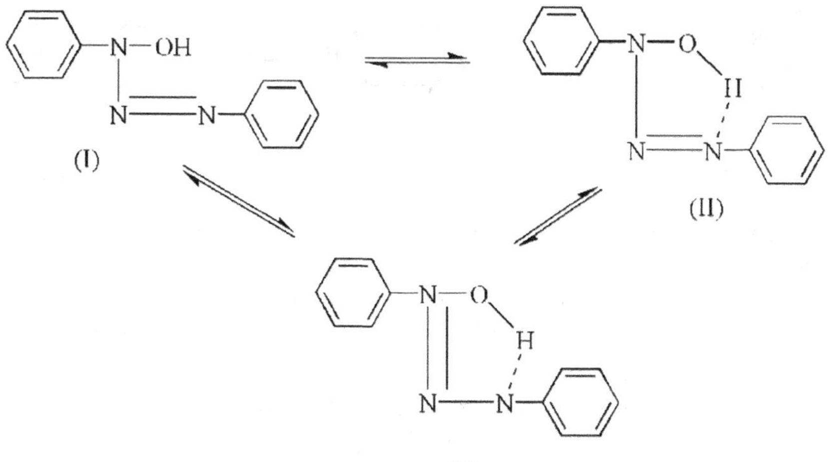

Figure 2.1 HDPT and other analogues exist in an intra-molecularly hydrogen bonded form (I) and its tautomers (II) and (III).

HDPT. It is thus suggested by the authors that HDPT and other analogues exist in an intra-molecularly hydrogen-bonded form (I) and its tautomeric form (II) and (III) (Figure 2.1).

The structural details described above have been supported by UV and IR spectral studies by the same authors. It is described in the paper that IR spectra in KBr showed bands at 3480 cm^{-1} and 3190 cm^{-1} for νOH and νNH, respectively. Further, the band at 3480 cm^{-1} did not exist in carbon tetrachloride, and Mitsuhasi [14] reported the position of a second band at 3250.6 cm^{-1} which has been established to be νNH through studies on ^{15}N (nitrogen remote from oxygen) in labeled HDPT. The existence of tautomeric forms is further strengthened due to the possibility of migration of the H atom via the formation of an intermediate chelate (IV).

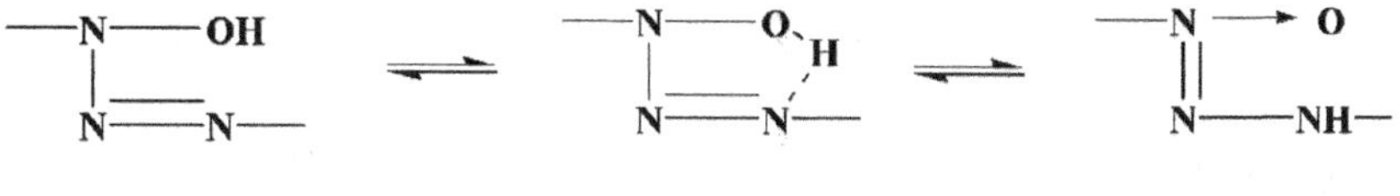

(IV)

The oxygen atom in the hydroxytriazene group is an electron donor, increasing the negative charge on the nitrogen atom, which increases the affinity of the H atom for nitrogen. Due to the greater electronegativity of oxygen compared to nitrogen a complete transfer of hydrogen to nitrogen does not occur. All these facts support a hydroxytriazene structure. The structure established can be represented as follows.

(V)   (VI)   (VII)

The strengthening of the N–N and N–O bonds is facilitated through the delocalization of electron density in the ring which equalizes them. Further, by this the π electron cloud of the molecule increases in symmetry. Thus a hydroxytriazene structure is predominant, as evidenced.

2.3.1 Tautomerism and hydrogen bonding in 3-hydroxy-1,3-diphenyltriazenes (HDPT)

Two leading publications on the tautomerism and association in hydroxytriazenes are available. On the basis of UV absorption studies, Calozolari and Furlani [11] suggested two possibilities for the structure of HPDT: (i) it exists in a single form, or (ii) two tautomeric forms have the same spectra. Purohit et al. suggested the existence of a hydrogen-bonded structure and its tautomeric forms based on spectral studies. The same conclusion was restated on the basis of the dissociation constant value of HDPT, pK = 11.41. The following resonating forms have been assigned.

(III)

(VII)

(IV)

(VIII)

(V)

(IX)

(VI)

(X)

Structures (I) to (VIII) show that charge on the anion is distributed over two benzene rings, making HDPT a stronger acid than phenol. However, the pK values indicate a contrary picture. A hydrogen-bonded structure and its tautomeric forms are suggestive of the suppression of acid character as shown in Figure 2.1 (II and III).

Further, on the basis of the spectrophotometric determination of HDPT [15], a hydroxytriazenic structure has been further evidenced.

References

1. (a) Bamberger, E. Ueber die Einwirkung des nitrosobezole auf Amidoverbin. *Ber.* 1895, 28, 104.
   (b) Bamberger, E., and Renauld, E. Ueber alphyllirte und alkylirte hydroxylamine. *Ber.* 1897, 30, 2280–2285.
   (c) Bamberger, E., and Busdorf, W. Ueber die Einwirkung von Nitrosbenzol auf aromatische hydrazine *Ber.* 1900, 33, 3510.
2. Sogani, N. C., and Bhattacharya, S. C. Preparation and study of hydroxytriazenes as analytical reagents Part I. *J. Indian. Chem. Soc.* 1960, 37, 531.
3. Mitsuhasi, T., and Simamura, O. Formation of 1,3-diaryltriazene 1-oxides by oxidation of diazoaminobenzenes with peroxybenzoic acid: Isomer ratios, kinetics, and mechanism. *J. Chem. Soc. (London)* 1970, 705.
4. Purohit, D. N. Spot test for hydroxytriazenes. *Analytica Chimica Acta* 1973, 63, 493.
5. Purohit, D. N. Spot test detection of hydroxytriazenes with dimethylaniline and hydrochloric acid. *Clencia E Cultura* 1986, 38, 1059.
6. Purohit, D. N. A new spot test for hydroxytriazenes with sulphuric acid. *Bol. Soc. Quin. (Peru)* 1986, 52, 224.
7. Purohit, D. N. Spot test for hydroxytriazenes. *Analytica Chimica Acta* 1973, 66, 463.
8. Katritzky, A. R. and Monro, A. M. N-oxides and related compounds. Part X. The hydrogenation of some pyridine 1-oxides. *Journal of Chemical Society,* Issue 0, 1958, v1263.
9. Purohit, D. N. Hydrogen bonding and tautomerism in 3-hydroxy-1,3-diphenyltriazene: Infrared studies. *Spectrochimica Acta* 1985, 41A, 873.
10. Purohit, D. N. Infra-red spectral studies on association of hydroxytriazenes. *Proc. Nat. Acad. Sci. India* 1985, III, 204.
11. Calzolari, C., and Furlani, A. D. Electrochemical reduction and polarographic behavior of 3-hydroxy-1,3-diphenyltriazene. *Gazz. Chem. Ital.* 1957, 87, 862.
12. Purohit, D. N., Dugar, S. M., and Sogani, N. C. Effect of substitution on the stabilities of copper chelates of hydroxytriazenes. *Z. Naturforsct* 1965, B-20, 853.
13. Purohit, D. N., and Sogani, N. C. Spectrophotometric study of iron(III) chelates of N-acetyl-N-phenylhydroxylamine. *J. Indian Chem. Soc.* 1964, 41, 160.
14. Mitsuhasi, T., Osamura, Y., and Simamura, O. The structure of so-called diazohydroxyaminobenzenes. *Tetrahedron Lelfers* 1965, 53, 3530.
15. Purohit, D. N., and Goswami, A. K. Spectrophotometric micro determination of hydroxytriazenes. *Talanta* 1973, 20, 689.

chapter three

Medicinal applications of hydroxytriazenes

3.1 Introduction

Historically, it is interesting to note that some 19th-century pharmacists working in their apothecary laboratories were the first to develop naturally occurring drugs. Rudolf Buchheim, a German pharmacologist, is known for his pioneering work in experimental pharmacology. He was the one who turned pharmacology from an empirical study of medicine into an exact science. He introduced the bioassay and created a methodology to develop chemical substances as drugs. He proposed that the mission of pharmacology was to establish the active ingredients within the material drugs and to find the mode of action as well as prepare them synthetically for better efficacy. In modern times, medicinal chemistry has developed as a chemistry-based discipline incorporating aspects of biological, medical, and pharmaceutical science. It is central to drug discovery and development. The roots of medicinal chemistry lie in a number of branches of chemistry and biology. These are essential parts of the process of drug development and are interwoven branches. The branch of medicinal chemistry is focused to the discovery and development of new synthetic or natural products for treating diseases. Although the majority of this is covered by new organic compounds, inorganic compounds have a significant share in therapy. The synthetic strategy for developing new molecules has gone beyond traditional methods to biotechnology which uses cells' protocols for synthesizing new compounds. The chemical structures of new compounds with trends in their biological behavior have opened up a new area of hypothesizing drug action and the simulated or probable mechanism of action. The new field of computer-aided drug design (CADD) also plays a contributory role in the development of new molecules. Medicinal chemistry has become an interdisciplinary branch of drug development using areas such as biology, CADD, X-ray crystallography, legal aspects of regulatory guidelines, process chemistry, and pharmacokinetics to name a few.

3.2 Triazenes and medicinal chemistry

An excellent research paper by Stevens and Newlands [1] shows the importance of triazenes and triazeies in medicinal chemistry as anticancer drugs. The paper describes the discovery of temozolomide as the first symbiotic fusion of two cultures, that is, the laboratory sciences of chemistry and pharmacology and on the other hand the clinical sciences which ultimately test the efficacy of drugs in patients. The ancestor molecules of temozolomide were synthesized in the early 1960s. Decarbazene and temocolomide [2,3] are two members of the triazene class used in the treatment of metastatic melanomas, soft tissue, and Hodgkin and non-Hodgkin lymphoma. The antitumor activities of the drugs are supposed to be due to three adjacent nitrogen atoms [4–8]. The lipophilicity of temozolomide facilitates it to cross the blood–brain barrier and makes it a first-line therapeutic drug for the treatment of metastatic brain tumors [9,10]. Further, a well-known diaryltriazene derivative is diminazene aceturate (Berenil) which is a salt of 1,3-bis-(4-amidinophenyl) triazene. Its capacity to bind to DNA was recognized long ago. The DNA binding occurs through complexation into the minor groove of AT-rich domains of double helices [12]. Diminazene aceturate can also bind to RNA as well as DNA duplex showing both intercalation as well as minor groove binding. The low toxicity and good antitumor effects make aromatic triazenes good drug candidates [13–16].

Triazenes have received attention lately in the search for potential HIV-1 inhibitors [17]. They act as chemotherapeutics for a variety of tumors such as brain, melanoma lymphoma, and sarcoma [18–22].

3.3   *Theoretical drug design*

Theoretical aspects of drug design: Producing a new drug is an expensive process which is not only time-consuming but is always subject to extensive regulation. The simplest definition of a drug is a chemical or biological substance having some kind of physiological or biochemical effect on our body. The drug may be a single molecule or a mixture of many components. Although the effect of a drug is intended to be beneficial, it may cause some harmful side effects in some people. Every drug interacts with a specific target in the body with the objective to modify its activity, and this often results in a therapeutic effect. The targets are usually proteins, or small regions of RNA or DNA. The drugs either stimulate or block the activity of their targets.

The development of a new drug is a complex, cumbersome, and expensive process. It may take several years and a large amount of money to develop a drug. The process from initiation to concept, safety, and efficacy take years and a huge amount of money, before a drug reaches the hospital or market. It may take 2–4 years of preclinical development, a number of years of clinical development, and additional time with the regulatory authorities before it is launched as a drug.

In the past though, some drugs were discovered by accident; now a more defined and systematic approach is used. Scientists test thousands of potential targets with several different chemical compounds to identify new drugs. Another approach is rational drug design involving designing and synthesizing compounds based on the known structure of a specific target. High-throughput screening may identify several potential lead components, out of which many will be eliminated at the first round of testing. Rational drug design uses computer-based modeling to achieve this specificity. In the second stage the results of preclinical testing are used to determine the best formulation of a drug intended for clinical use. The clinical development is then done in five phases, O, I, II, III, and IV. Somehow by the end of the clinical development phase most of the new drugs are eliminated on the grounds of safety and effectiveness. One or two compounds are submitted as new drugs to the regulatory body. After approval by these bodies pharmaceutical companies have a short period with exclusive rights to the market the drug before any other company may market the drug. In phase IV, after finally launching a drug, new side effect risk factors not recorded previously still may be identified. This is part of the confirmed monitoring of the drug in the target patients. To review briefly, Figure 3.1 shows the entire process of drug development.

A more accurate term for drug design is ligand design, that is the designing of a molecule that binds to its target tightly [23]. A good success rate is achieved by designing techniques to predict the binding affinity of ligands, yet there is dearth of tools to predict many other properties

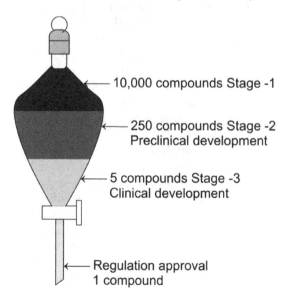

Figure 3.1 Drug discovery process.

such a bioavailability, side effects, and metabolic half-life. These are issues addressed by in vitro experiment supported by computational methods used in early drug discovery. Some of the available tools for theoretical designing of drug are

(1) Prediction of activity spectra of substances (PASS)
(2) Adsorption, distribution, metabolism, and excretion (ADME) and toxicological profile [24]

3.3.1   Prediction of activity spectra of substance (PASS)

PASS predicts, on the basis of the structure of a compound, over 4000 kinds of biological activities such as pharmacological effects, action, mechanisms, toxic and adverse effects, interactions with metabolic enzymes, transportation, and influence on gene expression to name a few. The prediction of all these activities is available online on the basis of the structural formula of the compound. This makes it an extremely useful tool, since one can predict the activity profile of even a virtual structure without synthesizing it. Thus, PASS predicts about 4000 kinds of activity simultaneously with a mean accuracy of prediction ~95% on the basis of the structural formula only. The application is extremely useful in the theoretical design of a drug since it reveals new effects and suggests a probable mechanism of action. The most promising compounds can be selected for further high-throughput screening, and finally the most relevant screens for particular compounds are found theoretically. It seems very simple and the success rate of the prediction still depends on a number of factors. It is helpful to those who are interested in designing a drug. A simplified use of PASS is illustrated using an example of the biological activity spectrum of one of the hydroxytriazenes whose experimental validation has been done by Shilpa Agarwal et al. [25]. The detailed strategy of theoretical prediction using PASS and experimental validation of some of the predicted activities is described in the next section.

The biological activity spectrum of any chemical compound using PASS reflects the results of the interaction of the chemical compound with different biological entities. The qualitative expression of the activity is defined as yes/none, suggesting that the intrinsic property of a substance basically depends only on its structure and physicochemical characteristics. Although it's a gross generalization, it facilitates the possibility of combining information from many sources in the common training set. This is necessary as the information from one particular publication does not cover the various biological action of a compound. A very brief introduction of PASS and related terms is given here for reference purposes only [26,27]. Probability to be active (Pa)—predicts the chances that the compound studied belongs to the sub-class of active compounds, that is, it resembles the structure of molecules which are the most typical in a sub-set of activities in a training set, whereas probability to be inactive (Pi) describes the inactivity of the compounds. In general, compounds with Pa values of more than 0.3 (30%) can be expected to have the predicted activities and those with Pa < 0.3 may not possess the predicted activity. To demonstrate this, two tables are given for a hydroxytriazene used for the strategy of theoretical prediction and experimental validation of some of the predicted properties. A sample table for 1,3-diphenyl hydroxytriazene and predicted activities along with synthesized compounds and their predicted activities is given only to illustrate the application of PASS (Tables 3.1 and 3.2).

Summarily the structure and its variants can be screened before synthesis for the biological activity and out of several activities those having values of Pa > 0.3 can be experimentally validated. This planning can save a lot of resources such as chemicals as well as the time of the researcher. The authors' research group has exploited the best use of this. The actual medicinal application of the hydroxytriazenes is presented in the next chapter. However, the authors do not claim any success rate, but in our case the rate for anti-inflammatory and anti-radical properties has been excellent.

3.3.2   *The absorption, distribution, metabolism,*
excretion (ADME) and toxicity profiling

An excellent review entitled "ADMET In Silico Modelling: Towards Prediction Paradise?" written by Han Van de Waterbeemd and Giffors [28] describes the significance and need of ADMET. After 1990, poor pharmacokinetics and toxicity were important causes of failures in drug development. It was felt that these areas were of prime consideration at the earliest stage of drug discovery. On the other hand, the development of combinatorial chemistry as well as high-throughput screens further increased the number of compounds for which data were needed a priori. This has significantly increased the importance of ADMET in the drug discovery process. This challenge has been addressed using a clear and precise strategy. The testing of any drug before evaluation at the clinical stage is done well in advance. This includes testing of the drug metabolism pharmacokinetics or even toxicity. This is a new approach to combinatorial chemistry. It facilitates the synthesis of a large number and series of closely related libraries of chemicals. These libraries are then run

Table 3.1 PASS of 1,3-diphenyl-3-hydroxytriazenes

⊙ ○ Pa > Pi; ○ Pa > 0.3; ○ Pa > 0.7

P_a	P_i	Predicted activity
0.821	0.003	Thiosulfate dehydrogenase inhibitor
0.804	0.007	5-O-(4-coumaroyl)-D-quinate 3′-monooxygenase inhibitor
0.812	0.030	Aspulvinone dimethylallyltransferase inhibitor
0.796	0.021	Acrocylindropepsin inhibitor
0.796	0.021	Chymosin inhibitor
0.796	0.021	Saccharopepsin inhibitor
0.767	0.005	Corticosteroid side chain isomerase inhibitor
0.776	0.015	NADPH peroxidase inhibitor
0.768	0.009	Prolyl aminopeptidase inhibitor
0.756	0.002	AICAR transformylase inhibitor
0.765	0.020	Glycosylphosphatidylinositol phospholipase D inhibitor
0.744	0.009	Phospholipid-translocating ATPase inhibitor
0.732	0.008	Pterin deaminase inhibitor
0.736	0.013	Fragilysin inhibitor
0.733	0.018	Dehydro-L-gulonate decarboxylase inhibitor
0.736	0.027	Sugar phosphatase inhibitor
0.724	0.016	Pseudolysin inhibitor
0.728	0.021	Glutamyl endopeptidase II inhibitor
0.746	0.040	Testosterone 17-β-dehydrogenase (NADP+) inhibitor
0.707	0.007	S-alkylcysteine lyase inhibitor
0.718	0.019	Fusarinine-C ornithinesterase inhibitor
0.701	0.011	L-glutamate oxidase inhibitor
0.692	0.009	Aspartate-phenylpyruvate transaminase inhibitor
0.696	0.014	IgA-specific serine endopeptidase inhibitor
0.694	0.014	Bisphosphoglycerate phosphatase inhibitor
0.723	0.043	CYP2J substrate
0.738	0.059	Phobic disorders treatment
0.733	0.056	Ubiquinol-cytochrome-c reductase inhibitor
0.694	0.020	Carboxypeptidase Taq inhibitor
0.688	0.017	Phthalate 4,5-dioxygenase inhibitor
0.688	0.020	Complement factor D inhibitor
0.676	0.008	Aminobutyraldehyde dehydrogenase inhibitor
0.688	0.022	Ribulose-phosphate 3-epimerase inhibitor
0.686	0.026	Glucose oxidase inhibitor
0.681	0.021	Glutathione thiolesterase inhibitor
0.666	0.008	Glyoxylate reductase inhibitor

(Continued)

Table 3.1 (Continued) PASS of 1,3-diphenyl-3-hydroxytriazenes

⊙ ○ Pa > Pi; ○ Pa > 0.3; ○ Pa > 0.7

P_a	P_i	Predicted activity
0.660	0.004	Taurine-2-oxoglutarate transaminase inhibitor
0.659	0.003	Trimethylamine-N-oxide reductase inhibitor
0.693	0.037	Nicotinic α6β3β4α5 receptor antagonist
0.670	0.017	Cl-transporting ATPase inhibitor
0.666	0.014	Nucleoside oxidase (H_2O_2-forming) inhibitor
0.661	0.010	(S)-6-hydroxynicotine oxidase inhibitor
0.668	0.017	Electron-transferring flavoprotein dehydrogenase inhibitor
0.685	0.037	CYP2J2 substrate
0.689	0.042	Polyporopepsin inhibitor
0.660	0.018	Arylacetonitrilase inhibitor
0.677	0.037	Taurine dehydrogenase inhibitor
0.656	0.016	Chloride peroxidase inhibitor
0.657	0.018	Thioredoxin inhibitor
0.650	0.012	Fibrolase inhibitor
0.640	0.012	Chenodeoxycholoyltaurine hydrolase inhibitor
0.651	0.024	UDP-N-acetylglucosamine 4-epimerase inhibitor
0.637	0.012	Gamma-guanidinobutyraldehyde dehydrogenase inhibitor
0.640	0.016	3-Hydroxybenzoate 6-monooxygenase inhibitor
0.649	0.026	Glucan endo-1,6-β-glucosidase inhibitor
0.635	0.013	Cutinase inhibitor
0.634	0.016	(R)-6-hydroxynicotine oxidase inhibitor
0.640	0.022	Phosphatidylcholine-retinol O-acyltransferase inhibitor
0.632	0.020	Manganese peroxidase inhibitor
0.618	0.007	Plastoquinol-plastocyanin reductase inhibitor
0.628	0.017	Platelet aggregation stimulant
0.645	0.038	Nicotinic α2β2 receptor antagonist
0.631	0.025	Alkane 1-monooxygenase inhibitor
0.663	0.062	Membrane integrity agonist
0.638	0.036	Pro-opiomelanocortin converting enzyme inhibitor
0.622	0.022	Peroxidase inhibitor
0.606	0.006	4-Hydroxyglutamate transaminase inhibitor
0.603	0.010	Hyponitrite reductase inhibitor
0.613	0.021	Gluconate 5-dehydrogenase inhibitor
0.597	0.006	3-Hydroxybenzoate 4-monooxygenase inhibitor
0.614	0.026	Limulus clotting factor B inhibitor
0.624	0.038	Omptin inhibitor

(Continued)

Table 3.1 (Continued) PASS of 1,3-diphenyl-3-hydroxytriazenes

◉ ○ Pa > Pi; ○ Pa > 0.3; ○ Pa > 0.7

P_a	P_i	Predicted activity
0.602	0.017	Spermidine dehydrogenase inhibitor
0.603	0.018	Lysostaphin inhibitor
0.615	0.032	NADPH-cytochrome-c-2 reductase inhibitor
0.591	0.011	N-acylmannosamine kinase inhibitor
0.608	0.030	Venombin AB inhibitor
0.599	0.025	Polyamine-transporting ATPase inhibitor
0.586	0.014	Arylmalonate decarboxylase inhibitor
0.606	0.033	Alkylacetylglycerophosphatase inhibitor
0.609	0.037	Kidney function stimulant
0.586	0.016	N-formylmethionyl-peptidase inhibitor
0.582	0.011	Cyclohexyl-isocyanide hydratase inhibitor
0.602	0.032	Feruloyl esterase inhibitor
0.585	0.017	Ferredoxin-NAD+ reductase inhibitor
0.585	0.017	Naphthalene 1,2-dioxygenase inhibitor
0.580	0.012	Glucan 1,4-α-maltotetraohydrolase inhibitor
0.585	0.017	tRNA-pseudouridine synthase I inhibitor
0.579	0.013	N-methylhydantoinase (ATP-hydrolyzing) inhibitor
0.587	0.022	Antiviral (Picornavirus)
0.571	0.021	Glucan 1,4-α-maltotriohydrolase inhibitor
0.561	0.011	Bothrolysin inhibitor
0.584	0.034	2-Dehydropantoate 2-reductase inhibitor
0.579	0.030	Dimethylargininase inhibitor
0.583	0.034	Arginine 2-monooxygenase inhibitor
0.561	0.012	2-Haloacid dehalogenase inhibitor
0.577	0.030	Fatty-acyl-CoA synthase inhibitor
0.591	0.046	Alkenylglycerophosphocholine hydrolase inhibitor
0.579	0.037	Macrophage colony stimulating factor agonist
0.566	0.024	Phenol O-methyltransferase inhibitor
0.561	0.022	CYP2D16 substrate
0.600	0.061	Chlordecone reductase inhibitor
0.580	0.046	Acylcarnitine hydrolase inhibitor
0.545	0.013	Crotonoyl-[acyl-carrier-protein] hydratase inhibitor
0.541	0.009	CDK9/cyclin T1 inhibitor
0.588	0.057	Antiseborrheic
0.556	0.026	Pullulanase inhibitor
0.545	0.015	Opheline kinase inhibitor
0.545	0.015	Taurocyamine kinase inhibitor

(Continued)

Table 3.1 (Continued) PASS of 1,3-diphenyl-3-hydroxytriazenes

◉ ○ Pa > Pi; ○ Pa > 0.3; ○ Pa > 0.7

P_a	P_i	Predicted activity
0.555	0.027	Methylamine-glutamate N-methyltransferase inhibitor
0.557	0.029	Hydrogen dehydrogenase inhibitor
0.550	0.025	CYP2A8 substrate
0.543	0.018	NADH kinase inhibitor
0.550	0.027	Linoleate diol synthase inhibitor
0.547	0.026	Glutamine-phenylpyruvate transaminase inhibitor
0.542	0.022	Mannan endo-1,4-β-mannosidase inhibitor
0.527	0.011	Myeloblastin inhibitor
0.529	0.019	Cis-1,2-dihydro-1,2-dihydroxynaphthalene dehydrogenase inhibitor
0.529	0.021	Prostaglandin-A1 DELTA-isomerase inhibitor
0.518	0.011	Polyneuridine-aldehyde esterase inhibitor
0.516	0.011	Arylalkyl acylamidase inhibitor
0.513	0.011	Dimethylmaleate hydratase inhibitor
0.513	0.012	Catechol oxidase inhibitor
0.510	0.012	Horrilysin inhibitor
0.540	0.044	JAK2 expression inhibitor
0.512	0.016	Albendazole monooxygenase inhibitor
0.520	0.025	Amine dehydrogenase inhibitor
0.532	0.037	1,4-Lactonase inhibitor
0.536	0.042	Lysine 2,3-aminomutase inhibitor
0.532	0.038	Alopecia treatment
0.501	0.010	Camphor 1,2-monooxygenase inhibitor
0.507	0.017	Carbon monoxide dehydrogenase inhibitor
0.507	0.016	Opine dehydrogenase inhibitor
0.507	0.016	Glyoxylate oxidase inhibitor
0.509	0.019	Phosphopantothenoylcysteine decarboxylase inhibitor
0.517	0.027	All-*trans*-retinyl-palmitate hydrolase inhibitor
0.523	0.033	Trimethylamine-oxide aldolase inhibitor
0.527	0.039	Exoribonuclease II inhibitor
0.502	0.015	N-acyl-D-aspartate deacylase inhibitor
0.506	0.018	Sulfite oxidase inhibitor
0.521	0.036	Phosphatidylserine decarboxylase inhibitor
0.519	0.034	Formaldehyde transketolase inhibitor
0.534	0.051	GST A substrate
0.498	0.016	Methanol dehydrogenase inhibitor
0.517	0.036	Endopeptidase So inhibitor

(Continued)

Table 3.1 (Continued) PASS of 1,3-diphenyl-3-hydroxytriazenes

⊙   ○ Pa > Pi; ○ Pa > 0.3; ○ Pa > 0.7

P_a	P_i	Predicted activity
0.509	0.030	Mucinaminylserine mucinaminidase inhibitor
0.506	0.027	Cytochrome P450 stimulant
0.510	0.031	Acetylesterase inhibitor
0.524	0.045	2-Hydroxyquinoline 8-monooxygenase inhibitor
0.500	0.022	Carnitinamidase inhibitor
0.517	0.039	Fructose 5-dehydrogenase inhibitor
0.503	0.027	APOA1 expression enhancer
0.500	0.024	Thymidylate 5'-phosphatase inhibitor
0.488	0.012	2-Hydroxy-3-oxoadipate synthase inhibitor
0.499	0.027	Hydroxylamine oxidase inhibitor
0.506	0.037	4-Nitrophenol 2-monooxygenase inhibitor
0.497	0.029	Creatininase inhibitor
0.478	0.009	Indoleacetaldoxime dehydratase inhibitor
0.514	0.047	27-Hydroxycholesterol 7-α-monooxygenase inhibitor
0.481	0.017	Glycolate dehydrogenase inhibitor
0.495	0.031	CYP2F1 substrate
0.484	0.021	Laccase inhibitor
0.479	0.017	Aldehyde dehydrogenase (pyrroloquinoline-quinone) inhibitor
0.471	0.010	Anti-inflammatory, intestinal
0.470	0.010	N-acetyl-gamma-glutamyl-phosphate reductase inhibitor
0.489	0.031	Nicotine dehydrogenase inhibitor
0.472	0.014	Sorbitol-6-phosphate 2-dehydrogenase inhibitor
0.476	0.018	Shikimate O-hydroxycinnamoyltransferase inhibitor
0.480	0.023	Alanine-tRNA ligase inhibitor
0.491	0.036	ADP-thymidine kinase inhibitor
0.509	0.054	Glucan endo-1,3-β-D-glucosidase inhibitor
0.490	0.035	CYP2B5 substrate
0.486	0.035	Histidine N-acetyltransferase inhibitor
0.465	0.015	2,3-Dihydroxyindole 2,3-dioxygenase inhibitor
0.474	0.025	N-benzyloxycarbonylglycine hydrolase inhibitor
0.477	0.029	Urethanase inhibitor
0.458	0.010	2,4-Diaminopentanoate dehydrogenase inhibitor
0.458	0.010	3-Hydroxybutyryl-CoA dehydrogenase inhibitor
0.458	0.010	Lysine 6-dehydrogenase inhibitor
0.463	0.016	Tryptophanamidase inhibitor
0.471	0.024	IgA-specific metalloendopeptidase inhibitor

(Continued)

Table 3.1 (Continued) PASS of 1,3-diphenyl-3-hydroxytriazenes

⊙ ○ Pa > Pi; ○ Pa > 0.3; ○ Pa > 0.7

P_a	P_i	Predicted activity
0.455	0.009	Gallate decarboxylase inhibitor
0.463	0.017	Meprin B inhibitor
0.459	0.014	CDP-4-dehydro-6-deoxyglucose reductase inhibitor
0.464	0.020	Gluconolactonase inhibitor
0.461	0.017	Quinoprotein glucose dehydrogenase inhibitor
0.468	0.027	2-Hydroxymuconate-semialdehyde hydrolase inhibitor
0.443	0.003	Azobenzene reductase inhibitor
0.446	0.006	CYP2C3 substrate
0.462	0.022	Cyanoalanine nitrilase inhibitor
0.471	0.032	Na+-transporting two-sector ATPase inhibitor
0.484	0.044	Erythropoiesis stimulant
0.466	0.027	Aspartate-ammonia ligase inhibitor
0.472	0.034	S-formylglutathione hydrolase inhibitor
0.452	0.015	Benzaldehyde dehydrogenase (NADP+) inhibitor
0.450	0.015	Linoleoyl-CoA desaturase inhibitor
0.457	0.022	Nicotinate dehydrogenase inhibitor
0.485	0.051	Leukopoiesis stimulant
0.448	0.014	Myosin ATPase inhibitor
0.452	0.018	Tetrahydroxynaphthalene reductase inhibitor
0.450	0.017	2-Oxoaldehyde dehydrogenase (NADP+) inhibitor
0.472	0.040	Octopamine antagonist
0.488	0.057	Preneoplastic conditions treatment
0.497	0.066	Lysase inhibitor
0.457	0.027	Nitrite reductase (NO-forming) inhibitor
0.467	0.036	Arylsulfate sulfotransferase inhibitor
0.462	0.033	Tpr proteinase (Porphyromonas gingivalis) inhibitor
0.442	0.015	N-hydroxy-2-acetamidofluorene reductase inhibitor
0.452	0.026	Long chain aldehyde dehydrogenase inhibitor
0.446	0.020	Clavaminate synthase inhibitor
0.442	0.016	Alkylglycerone-phosphate synthase inhibitor
0.440	0.015	Ferredoxin hydrogenase inhibitor
0.433	0.008	3-Chloro-D-alanine dehydrochlorinase inhibitor
0.432	0.008	Pantoate 4-dehydrogenase inhibitor
0.448	0.024	EIF4E expression inhibitor
0.438	0.014	Snapalysin inhibitor
0.431	0.009	X-Pro dipeptidase inhibitor
0.461	0.039	CYP2A4 substrate

(Continued)

Table 3.1 (Continued) PASS of 1,3-diphenyl-3-hydroxytriazenes

⊙ ○ Pa > Pi; ○ Pa > 0.3; ○ Pa > 0.7

P_a	P_i	Predicted activity
0.456	0.036	CYP2D15 substrate
0.431	0.011	Quercetin 2,3-dioxygenase inhibitor
0.471	0.051	Nitrate reductase (cytochrome) inhibitor
0.433	0.013	Cyclopentanone monooxygenase inhibitor
0.434	0.015	Styrene-oxide isomerase inhibitor
0.444	0.026	Poly(β-D-mannuronate) lyase inhibitor
0.431	0.013	Bisphosphoglycerate mutase inhibitor
0.430	0.013	CYP2A3 substrate
0.441	0.024	Fructan β-fructosidase inhibitor
0.432	0.015	6-Carboxyhexanoate-CoA ligase inhibitor
0.432	0.015	Biotin-CoA ligase inhibitor
0.432	0.015	Homoaconitate hydratase inhibitor
0.432	0.015	Triacetate-lactonase inhibitor
0.432	0.015	2,6-Dihydroxypyridine 3-monooxygenase inhibitor
0.440	0.024	Salicylate 1-monooxygenase inhibitor
0.432	0.016	Xenobiotic-transporting ATPase inhibitor
0.448	0.034	Methylumbelliferyl-acetate deacetylase inhibitor
0.437	0.023	Poly(α-L-guluronate) lyase inhibitor
0.462	0.050	Biotinidase inhibitor
0.535	0.127	Membrane permeability inhibitor
0.429	0.022	3-Cyanoalanine hydratase inhibitor
0.451	0.044	Alkenylglycerophosphoethanolamine hydrolase inhibitor
0.437	0.031	Thiol oxidase inhibitor
0.426	0.020	Anthranilate-CoA ligase inhibitor
0.440	0.034	Aspergillopepsin I inhibitor
0.455	0.052	Sulfur reductase inhibitor
0.425	0.022	Transketolase inhibitor
0.419	0.016	N-carbamoyl-L amino acid hydrolase inhibitor
0.425	0.023	Glutarate-semialdehyde dehydrogenase inhibitor
0.440	0.039	Leukotriene-B4 20-monooxygenase inhibitor
0.432	0.031	3-Phytase inhibitor
0.418	0.018	Botulin neurotoxin A light chain inhibitor
0.416	0.018	Aryldialkylphosphatase inhibitor
0.417	0.019	CDP-diacylglycerol-glycerol-3-phosphate 3-phosphatidyltransferase inhibitor
0.424	0.028	Sulfite reductase inhibitor
0.406	0.011	Thiamine pyridinylase inhibitor

(Continued)

Table 3.1 (Continued) PASS of 1,3-diphenyl-3-hydroxytriazenes

⊙ ○ Pa > Pi; ○ Pa > 0.3; ○ Pa > 0.7

P_a	P_i	Predicted activity
0.416	0.021	D-alanine 2-hydroxymethyltransferase inhibitor
0.434	0.039	Mitochondrial processing peptidase inhibitor
0.406	0.012	Lombricine kinase inhibitor
0.433	0.039	Simian immunodeficiency virus proteinase inhibitor
0.421	0.027	Coccolysin inhibitor
0.426	0.033	Pitrilysin inhibitor
0.429	0.035	Glycerol-3-phosphate oxidase inhibitor
0.406	0.013	Tyrosine 3 hydroxylase inhibitor
0.437	0.046	Centromere associated protein inhibitor
0.393	0.003	Antiprotozoal (Babesia)
0.409	0.020	γ-D-glutamyl-meso-diaminopimelate peptidase inhibitor
0.471	0.082	Protein-disulfide reductase (glutathione) inhibitor
0.441	0.052	Superoxide dismutase inhibitor
0.428	0.041	Limulus clotting factor C inhibitor
0.432	0.046	N-acetylneuraminate 7-O(or 9-O)-acetyltransferase inhibitor
0.443	0.057	Sphinganine kinase inhibitor
0.451	0.065	5-Hydroxytryptamine uptake stimulant
0.407	0.021	Leucolysin inhibitor
0.410	0.026	Hydroxylamine reductase (NADH) inhibitor
0.399	0.014	Indolepyruvate C-methyltransferase inhibitor
0.424	0.040	(R)-Pantolactone dehydrogenase (flavin) inhibitor
0.405	0.021	Mannitol-1-phosphatase inhibitor
0.403	0.019	Allyl-alcohol dehydrogenase inhibitor
0.411	0.027	Cholestanetriol 26-monooxygenase inhibitor
0.399	0.015	3-Carboxyethylcatechol 2,3-dioxygenase inhibitor
0.420	0.037	Leukopoiesis inhibitor
0.393	0.010	(R,R)-Butanediol dehydrogenase inhibitor
0.413	0.030	Threonine aldolase inhibitor
0.421	0.039	Peptide-N4-(N-acetyl-β-glucosaminyl)asparagine amidase inhibitor
0.427	0.046	Acetylgalactosaminyl-O-glycosyl-glycoprotein β-1,3-N-acetylglucosaminyltransferase inhibitor
0.395	0.015	Gly-X carboxypeptidase inhibitor
0.403	0.026	Di-trans, poly-cis-decaprenylcistransferase inhibitor
0.390	0.013	3-Oxoadipate enol-lactonase inhibitor
0.475	0.098	Calcium channel (voltage-sensitive) activator

(*Continued*)

Table 3.1 (Continued) PASS of 1,3-diphenyl-3-hydroxytriazenes

◉ ◯ Pa > Pi; ◯ Pa > 0.3; ◯ Pa > 0.7

P_a	P_i	Predicted activity
0.418	0.040	Phosphoinositide 5-phosphatase inhibitor
0.407	0.031	Undecaprenyldiphospho-muramoylpentapeptide β-N-acetylglucosaminyltransferase inhibitor
0.391	0.015	Creatinine deaminase inhibitor
0.392	0.016	Sweetener
0.415	0.039	Dolichyl-diphosphooligosaccharide-protein glycotransferase inhibitor
0.413	0.037	Alcohol dehydrogenase (acceptor) inhibitor
0.393	0.019	4-Chlorophenylacetate 3,4-dioxygenase inhibitor
0.387	0.013	(S)-3-hydroxyacid ester dehydrogenase inhibitor
0.411	0.039	Malate oxidase inhibitor
0.405	0.033	Mannose isomerase inhibitor
0.393	0.022	Glycerol 2-dehydrogenase (NADP+) inhibitor
0.397	0.027	Xylan endo-1,3-β-xylosidase inhibitor
0.395	0.026	Uroporphyrinogen-III synthase inhibitor
0.424	0.055	Methylenetetrahydrofolate reductase (NADPH) inhibitor
0.393	0.026	DNA-3-methyladenine glycosylase I inhibitor
0.370	0.004	Quisqualate antagonist
0.389	0.023	Oryzin inhibitor
0.396	0.030	Histidinol-phosphatase inhibitor
0.391	0.026	2,3,4,5-Tetrahydropyridine-2,6-dicarboxylate N-succinyltransferase inhibitor
0.375	0.011	CYP4A2 substrate
0.386	0.023	GABA C receptor agonist
0.381	0.019	Phosphoenolpyruvate mutase inhibitor
0.395	0.033	Pyruvate decarboxylase inhibitor
0.385	0.023	Ornithine cyclodeaminase inhibitor
0.384	0.023	Aminomuconate-semialdehyde dehydrogenase inhibitor
0.398	0.038	Chitosanase inhibitor
0.408	0.048	RNA-directed RNA polymerase inhibitor
0.408	0.049	Peptidoglycan glycosyltransferase inhibitor
0.411	0.053	Fucosterol-epoxide lyase inhibitor
0.383	0.025	Magnesium-protoporphyrin IX monomethyl ester (oxidative) cyclase inhibitor
0.407	0.051	FMO1 substrate
0.416	0.061	5-Hydroxytryptamine release inhibitor
0.397	0.044	4-Hydroxyproline epimerase inhibitor
0.402	0.049	N-hydroxyarylamine O-acetyltransferase inhibitor

(Continued)

Table 3.1 (Continued) PASS of 1,3-diphenyl-3-hydroxytriazenes

⊙ ○ Pa > Pi; ○ Pa > 0.3; ○ Pa > 0.7

P_a	P_i	Predicted activity
0.366	0.015	Pyruvate dehydrogenase (cytochrome) inhibitor
0.366	0.015	*trans*-2-enoyl-CoA reductase (NAD+) inhibitor
0.402	0.052	Cyclohexanone monooxygenase inhibitor
0.367	0.018	CYP2C29 substrate
0.373	0.024	Acetylornithine deacetylase inhibitor
0.389	0.040	P-benzoquinone reductase (NADPH) inhibitor
0.354	0.005	Aromatic-hydroxylamine O-acetyltransferase inhibitor
0.363	0.015	Undecaprenyl-diphosphatase inhibitor
0.412	0.064	Apyrase inhibitor
0.354	0.007	K(Ca) 3.1 channel activator
0.354	0.007	Potassium channel intermediate conductance Ca-activated activator
0.377	0.029	Glutathione dehydrogenase (ascorbate) inhibitor
0.371	0.024	Glycine dehydrogenase (decarboxylating) inhibitor
0.415	0.068	Lipoprotein lipase inhibitor
0.373	0.027	Chaperonin ATPase inhibitor
0.445	0.099	CYP2C12 substrate
0.424	0.079	MAP kinase stimulant
0.360	0.015	Maltose-transporting ATPase inhibitor
0.365	0.021	6-Pyruvoyltetrahydropterin synthase inhibitor
0.359	0.017	tRNA nucleotidyltransferase inhibitor
0.424	0.082	Oxygen scavenger
0.393	0.051	GABA aminotransferase inhibitor
0.452	0.110	Platelet adhesion inhibitor
0.383	0.042	NADPH-ferrihemoprotein reductase inhibitor
0.382	0.041	4-Methoxybenzoate monooxygenase (O-demethylating) inhibitor
0.417	0.077	Pin1 inhibitor
0.404	0.064	Immunosuppressant
0.403	0.064	Adenomatous polyposis treatment
0.384	0.046	Sulfite dehydrogenase inhibitor
0.381	0.043	Adenylyl-sulfate reductase inhibitor
0.360	0.022	Trimethylamine dehydrogenase inhibitor
0.353	0.016	Penicillin amidase inhibitor
0.373	0.036	Cyclomaltodextrinase inhibitor
0.373	0.037	Glycerol-3-phosphate dehydrogenase inhibitor
0.349	0.013	3-Hydroxyphenylacetate 6-hydroxylase inhibitor

(Continued)

Table 3.1 (Continued) PASS of 1,3-diphenyl-3-hydroxytriazenes

⊙ ○ Pa > Pi; ○ Pa > 0.3; ○ Pa > 0.7

P_a	P_i	Predicted activity
0.360	0.024	Ubiquitin thiolesterase inhibitor
0.344	0.009	Ligase inhibitor
0.357	0.023	D-amino acid dehydrogenase inhibitor
0.379	0.045	D-lactaldehyde dehydrogenase inhibitor
0.416	0.082	Neurotransmitter antagonist
0.343	0.009	Tropinesterase inhibitor
0.366	0.033	Succinate-semialdehyde dehydrogenase [NAD(P)+] inhibitor
0.348	0.016	N-acetylneuraminate synthase inhibitor
0.354	0.023	3-Demethylubiquinone-9 3-O-methyltransferase inhibitor
0.352	0.022	Prunasin β-glucosidase inhibitor
0.361	0.031	Yeast ribonuclease inhibitor
0.356	0.027	Urease inhibitor
0.354	0.026	3-Hydroxy-4-oxoquinoline 2,4-dioxygenase inhibitor
0.352	0.024	Pseudouridylate synthase inhibitor
0.372	0.045	Isopenicillin-N epimerase inhibitor
0.348	0.022	Pappalysin-1 inhibitor
0.347	0.021	Aspergillopepsin II inhibitor
0.343	0.017	Aureolysin inhibitor
0.370	0.044	Peptide α-N-acetyltransferase inhibitor
0.425	0.100	Mucositis treatment
0.340	0.018	Methylaspartate ammonia lyase inhibitor
0.346	0.025	D-Xylulose reductase inhibitor
0.361	0.040	Venom exonuclease inhibitor
0.358	0.037	Antiviral (Poxvirus)
0.335	0.017	Phosphoenolpyruvate protein phosphotransferase inhibitor
0.346	0.029	Hyaluronic acid agonist
0.369	0.051	Retinoic acid metabolism inhibitor
0.384	0.066	Rubredoxin-NAD+ reductase inhibitor
0.339	0.022	Vomilenine glucosyltransferase inhibitor
0.356	0.039	Arylesterase inhibitor
0.334	0.018	Mannan endo-1,6-α-mannosidase inhibitor
0.355	0.040	Para amino benzoic acid antagonist
0.331	0.016	Glycerol-1-phosphatase inhibitor
0.343	0.028	Phenylalanine(histidine) transaminase inhibitor
0.343	0.028	Serine-pyruvate transaminase inhibitor
0.341	0.026	2-Oxoglutarate decarboxylase inhibitor
0.323	0.010	Pyridoxine 4-oxidase inhibitor

(Continued)

Table 3.1 (Continued) PASS of 1,3-diphenyl-3-hydroxytriazenes

◉ ○ Pa > Pi; ○ Pa > 0.3; ○ Pa > 0.7

P_a	P_i	Predicted activity
0.361	0.047	4-Hydroxymandelate oxidase inhibitor
0.378	0.065	NAD(P)+-arginine ADP-ribosyltransferase inhibitor
0.385	0.073	Pancreatic elastase inhibitor
0.430	0.119	Acetylcholine neuromuscular blocking agent
0.363	0.052	Cathepsin T inhibitor
0.330	0.020	NAD+ synthase (glutamine-hydrolyzing) inhibitor
0.338	0.028	Sclerosant
0.315	0.004	Cytochrome-b-5 reductase substrate
0.342	0.033	Thiopurine S-methyltransferase inhibitor
0.323	0.014	Anthranilate 3-monooxygenase (deaminating) inhibitor
0.375	0.067	Peptidyl-dipeptidase Dcp inhibitor
0.332	0.025	Lactaldehyde reductase inhibitor
0.326	0.018	Endo-1,3(4)-β-glucanase inhibitor
0.340	0.033	CYP2B10 substrate
0.330	0.024	Inulinase inhibitor
0.348	0.042	ATP phosphoribosyltransferase inhibitor
0.333	0.027	Rhamnulose-1-phosphate aldolase inhibitor
0.346	0.040	CYP2G1 substrate
0.346	0.041	CYP2B11 substrate
0.333	0.028	2,4-Dichlorophenol 6-monooxygenase inhibitor
0.332	0.030	L-Threonine 3-dehydrogenase inhibitor
0.374	0.072	Malate dehydrogenase (acceptor) inhibitor
0.355	0.053	Cytostatic
0.360	0.058	Leukotriene-C4 synthase inhibitor
0.336	0.034	GST M substrate
0.333	0.033	Glucuronate isomerase inhibitor
0.317	0.018	α-N-arabinofuranosidase inhibitor
0.345	0.046	CYP2C10 substrate
0.339	0.042	Glutamate-tRNA ligase inhibitor
0.316	0.019	*trans*-pentaprenyltranstransferase inhibitor
0.344	0.048	Loop diuretic
0.310	0.014	Alkene monooxygenase inhibitor
0.410	0.114	Ovulation inhibitor
0.306	0.011	D-cysteine desulfhydrase inhibitor
0.321	0.026	Guanidinoacetase inhibitor
0.306	0.013	Iduronate-2-sulfatase inhibitor
0.308	0.014	Allantoinase inhibitor

(Continued)

Table 3.1 (Continued) PASS of 1,3-diphenyl-3-hydroxytriazenes

⊙ ○ Pa > Pi; ○ Pa > 0.3; ○ Pa > 0.7

P_a	P_i	Predicted activity
0.312	0.019	Peptidylamidoglycolate lyase inhibitor
0.419	0.126	CYP3A2 substrate
0.324	0.033	Phenylacetate-CoA ligase inhibitor
0.311	0.020	N-carbamoyl-D-amino acid hydrolase inhibitor
0.350	0.060	Steroid N-acetylglucosaminyltransferase inhibitor
0.305	0.014	Ribonucleoside triphosphate reductase inhibitor
0.326	0.036	Procollagen N-endopeptidase inhibitor
0.302	0.013	Aldehyde ferredoxin oxidoreductase inhibitor
0.305	0.016	3-Aminobutyryl-CoA ammonia lyase inhibitor
0.340	0.051	X-methyl-His dipeptidase inhibitor
0.322	0.035	Glycine amidinotransferase inhibitor
0.301	0.014	2,2-Dialkylglycine decarboxylase (pyruvate) inhibitor
0.326	0.040	Carminative
0.329	0.042	Ethanolamine-phosphate cytidylyltransferase inhibitor
0.310	0.024	N-acyl-D amino acid deacylase inhibitor
0.304	0.020	α-glucuronidase inhibitor
0.326	0.041	Imidazoline receptor agonist
0.303	0.019	Antihelmintic (Fasciola)
0.333	0.050	4-Hydroxyphenylacetate 3-monooxygenase inhibitor
0.330	0.047	Phosphatidylcholine-sterol O-acyltransferase inhibitor
0.415	0.133	TP53 expression enhancer
0.316	0.034	TRPA1 agonist
0.322	0.040	Mandelate 4-monooxygenase inhibitor
0.312	0.030	Rhodotorulapepsin inhibitor
0.344	0.063	Chitinase inhibitor
0.379	0.099	Cyclic AMP agonist
0.313	0.034	α-N-acetylglucosaminidase inhibitor
0.310	0.032	Lactose synthase inhibitor
0.313	0.036	Dihydroxy-acid dehydratase inhibitor
0.322	0.045	NF-E2-related factor 2 stimulant
0.305	0.030	Polygalacturonase inhibitor
0.316	0.042	Lysyl oxidase inhibitor
0.317	0.043	Dipeptidase E inhibitor
0.321	0.047	Indanol dehydrogenase inhibitor
0.319	0.045	Endothelial growth factor antagonist
0.364	0.091	Ecdysone 20-monooxygenase inhibitor
0.327	0.054	Monodehydroascorbate reductase (NADH) inhibitor

(Continued)

Table 3.1 (Continued) PASS of 1,3-diphenyl-3-hydroxytriazenes

◉ ○ Pa > Pi; ○ Pa > 0.3; ○ Pa > 0.7

P_a	P_i	Predicted activity
0.328	0.055	FMO3 substrate
0.372	0.100	CYP3A1 substrate
0.368	0.096	Vasoprotector
0.386	0.114	Thromboxane B_2 antagonist
0.324	0.053	Hematopoietic inhibitor
0.331	0.061	Formate-dihydrofolate ligase inhibitor
0.304	0.036	Guanidinoacetate kinase inhibitor
0.332	0.064	CYP7 inhibitor
0.398	0.130	Gastrin inhibitor
0.307	0.040	N-acetyllactosamine synthase inhibitor
0.321	0.055	Dextranase inhibitor
0.416	0.151	Fibrinolytic
0.334	0.071	Antiprotozoal (Trypanosoma)
0.322	0.060	CYP2A5 substrate
0.361	0.099	CYP2A1 substrate
0.325	0.064	Nitrite reductase [NAD(P)H] inhibitor
0.317	0.056	Skeletal muscle relaxant
0.320	0.062	Aspergillus nuclease S1 inhibitor
0.334	0.076	CYP2C18 substrate
0.319	0.063	AR expression inhibitor
0.318	0.065	DNA-(apurinic or apyrimidinic site) lyase inhibitor
0.329	0.077	CYP19A1 expression inhibitor
0.304	0.052	DNA synthesis inhibitor
0.305	0.054	Cardiovascular analeptic
0.311	0.062	Gingipain K inhibitor
0.335	0.087	Caspase 3 stimulant
0.314	0.070	Opioid kappa 3 receptor antagonist
0.341	0.098	β glucuronidase inhibitor
0.311	0.069	Choline-phosphate cytidylyltransferase inhibitor
0.302	0.061	Antiprotozoal (Coccidial)
0.339	0.098	β-adrenergic receptor kinase inhibitor
0.339	0.098	G-protein-coupled receptor kinase inhibitor
0.329	0.090	CYP3A3 substrate
0.315	0.077	Antiviral (Adenovirus)
0.337	0.100	Antimyopathies
0.312	0.075	Enteropeptidase inhibitor
0.377	0.142	CYP2C8 inhibitor

(Continued)

Table 3.1 (Continued) PASS of 1,3-diphenyl-3-hydroxytriazenes

⊙ ○ Pa > Pi; ○ Pa > 0.3; ○ Pa > 0.7

P_a	P_i	Predicted activity
0.336	0.101	Dementia treatment
0.313	0.082	CYP2D2 inhibitor
0.330	0.104	CYP4A11 substrate
0.313	0.095	CYP2E1 inducer
0.347	0.134	Cytoprotectant
0.308	0.095	MMP9 expression inhibitor
0.317	0.107	CYP2C9 inducer
0.353	0.144	Antinociceptive
0.321	0.117	Caspase 8 stimulant
0.337	0.134	Diabetic neuropathy treatment
0.315	0.121	Nucleotide metabolism regulator
0.341	0.151	Antidyskinetic
0.307	0.122	Mycothiol-S-conjugate amidase inhibitor
0.304	0.128	Intermittent claudication treatment
0.357	0.197	Mucomembranous protector
0.312	0.153	Oxidoreductase inhibitor

through high-throughput screenings (HTS) to find new hits. More focused and relevant series are then designed and synthesized. This has increased the demand for ADMET data. Various HTS in vitro ADMET screens are available for use. This chapter simply provides a basic approach of various strategies used in the discovery and development of a new drug, and thus the details of absorption, distribution, metabolism, and excretion are beyond the scope of the book. There are research papers available for in vitro and in vivo assessments of new drug candidates based on this approach. The papers describe the assessment of ADME and pharmacokinetic properties during lead selection and optimization. The guidelines, benchmarks, and rules of thumb have been consolidated in the paper, and any researcher of the field can use this for planning research on future drugs of interest. The authors and group have used the two-pronged approach, one via pass prediction and the other via ADMET guidelines for hydroxytriazenes. A detailed description of the compounds and their medicinal application will be described in the next section.

3.4   *Medicinal applications of hydroxytriazenes*

The earliest reports on the medicinal application of hydroxytriazenes, particularly 3-hydroxytriazenes, date back to those of the 1960s by Borisov et al. [29]. The synthesis of 3-hydroxy-3-benzyl-1-ethoxyphenyltriazne

Table 3.2 Theoretical prediction of activities of synthesized hydroxytriazenes by PASS

Predicted activity	ASPT-1		ASPT-2		ASPT-3		ASPT-4		ASPT-5		ASPT-6		ASPT-7		ASPT-8		ASPT-9		ASPT-10		ASPT-11	
	Pa	Pi	Pa	Pi	Pa	Pi	Pa	Pi	Pa	Pi	Pa	Pi	Pa	Pi	Pa	Pi	Pa	Pi	Pa	Pi	Pa	Pi
NADPH peroxidase inhibitor	0.476	0.076	0.317	0.149	0.317	0.149	0.317	0.149	0.394	0.105	0.480	0.075	0.274	0.183	0.486	0.073	0.352	0.126	0.518	0.064	0.254	0.201
Cyclooxygenase inhibitor	0.351	0.005	0.309	0.008	0.381	0.005	0.381	0.005	0.373	0.005	0.389	0.004	0.331	0.006	0.166	0.044	0.167	0.044	0.200	0.030	0.136	0.058
Cyclooxygenase 1 inhibitor	195.0	028	0.262	0.019	247	020	0.247	0.020	0.215	0.025	0.252	0.020	0.213	0.025	0.089	0.066	0.092	0.064	0.123	0.046	–	–
Cyclooxygenase 2 inhibitor	0.311	0.005	0.277	0.006	0.337	0.004	0.337	0.004	0.320	0.004	0.345	0.004	0.282	0.005	0.142	0.031	0.143	0.031	0.178	0.020	0.122	0.041
Prostaglandin E1 antagonist	0.274	0.025	0.276	0.024	0.269	0.026	–	–	–	–	0.292	0.20	–	–	0.236	0.039	0.254	0.031	0.254	0.031	0.253	0.031
Prostaglandin antagonist	0.044	0.042	0.044	0.041	0.045	0.039	0.045	0.039	0.047	0.036	0.054	0.028	0.071	0.017	–	–	–	–	–	–	–	–
Prostaglandin F2 alpha antagonist	0.073	0.004	0.069	0.005	0.069	0.005	0.069	0.005	0.070	0.005	0.072	0.005	0.063	0.007	0.068	0.005	0.068	0.005	0.069	0.005	0.067	0.006
Leukotriene E4 antagonist	0.163	0.009	0.163	0.009	0.158	0.010	0.158	0.010	0.133	0.020	0.140	0.016	0.153	0.011	0.146	0.014	0.149	0.013	0.148	0.013	0.153	0.011
Non-steroidal anti-inflammatory agent	0.173	0.128	0.200	0.100	0.182	0.118	0.182	0.118	0.196	0.104	0.203	0.098	–	–	–	–	–	–	–	–	–	–
5-Lipoxygenase inhibitor	–	–	–	–	–	–	–	–	–	–	–	–	–	–	0.075	0.046	0.078	0.043	0.090	0.032	0.154	0.012
Lipoxygenase inhibitor	0.098	0.064	0.106	0.056	0.098	0.063	0.098	0.063	0.090	0.073	0.090	0.073	0.087	0.078	–	–	0.145	0.031	0.171	0.024	0.260	0.010
15-Lipoxygenase inhibitor	0.071	0.058	0.078	0.045	0.073	0.055	0.073	0.055	0.076	0.048	0.090	0.073	0.067	0.065	–	–	–	–	–	–	–	–
Thromboxane A2 antagonist	0.067	0.043	0.075	0.030	0.072	0.034	0.072	0.034	0.070	0.037	0.068	0.040	0.070	0.037	–	–	–	–	0.063	0.051	0.060	0.058
Thromboxane B2 antagonist	0.302	0.207	0.298	0.211	0.268	0.250	0.268	0.250	–	–	–	–	0.394	0.107	–	–	–	–	–	–	–	–

(Continued)

Table 3.2 (Continued) Theoretical prediction of activities of synthesized hydroxytriazenes by PASS

Predicted activity	ASPT-1		ASPT-2		ASPT-3		ASPT-4		ASPT-5		ASPT-6		ASPT-7		ASPT-8		ASPT-9		ASPT-10		ASPT-11	
	Pa	Pi	Pa	Pi	Pa	Pi	Pa	Pi	Pa	Pi	Pa	Pi	Pa	Pi	Pa	Pi	Pa	Pi	Pa	Pi	Pa	Pi
Thromboxane antagonist	0.074	0.064	0.081	0.052	0.078	0.057	0.078	0.057	0.086	0.043	0.085	0.045	0.090	0.038	–	–	–	–	–	–	–	–
Sulfonylureas	0.219	0.004	0.218	0.004	0.242	0.004	0.242	0.004	0.235	0.004	0.217	0.004	0.144	0.010	0.186	0.005	0.206	0.005	0.203	0.005	0.206	0.005
Pyruvate dehydrogenase kinase inhibitor	–	–	0.111	0.039	0.110	0.040	0.110	0.040	0.117	0.031	0.119	0.029	0.109	0.043	0.009	0.058	0.107	0.045	0.109	0.042	0.095	0.067
Fructose-2,6-bisphosphate 2-phosphatase inhibitor	0.139	0.049	0.125	0.060	0.143	0.047	0.143	0.047	–	–	0.096	0.094	0.171	0.033	0.105	0.080	0.125	0.060	0.116	0.069	0.112	0.072
Glycerol-3-phosphate oxidase inhibitor	0.306	0.128	0.266	0.186	0.252	0.211	0.252	0.211	0.319	0.112	0.341	0.091	–	–	0.2490	0.234	0.293	0.145	0.321	0.110	–	–
Glutathione reductase	–	–	–	–	0.129	0.126	–	–	–	–	0.129	0.127	–	–	–	–	–	–	–	–	–	–
Antineoplastic (small cell lung cancer)	–	–	0.209	0.169	–	–	–	–	0.203	0.183	0.203	0.185	–	–	–	–	–	–	–	–	–	–
Antineoplastic (pancreatic cancer)	–	–	–	–	–	–	–	–	–	–	–	–	–	–	0.391	0.019	0.396	0.017	0.424	0.012	0.367	0.025
Antineoplastic (sarcoma)	–	–	–	–	–	–	–	–	–	–	–	–	–	–	0.286	0.020	0.302	0.017	0.322	0.014	0.266	0.024
Antineoplastic (brain cancer)	–	–	–	–	–	–	–	–	–	–	–	–	–	–	–	–	0.183	0.162	0.188	0.153	–	–

(Continued)

Table 3.2 (Continued) Theoretical prediction of activities of synthesized hydroxytriazenes by PASS

Predicted activity	ASPT-1		ASPT-2		ASPT-3		ASPT-4		ASPT-5		ASPT-6		ASPT-7		ASPT-8		ASPT-9		ASPT-10		ASPT-11	
	Pa	Pi	Pa	Pi	Pa	Pi	Pa	Pi	Pa	Pi	Pa	Pi	Pa	Pi	Pa	Pi	Pa	Pi	Pa	Pi	Pa	Pi
Antineoplastic (non-small cell lung cancer)	–	–	–	–	–	–	–	–	–	–	–	–	–	–	–	–	0.160	0.140	–	–	–	–

Compound code/name

(i) 3-hydroxy-3-phenyl-1-(4-acetylsulfonyl)phenyltriazene (ASPT-1)

(ii) 3-hydroxy-3-(2-methylphenyl)-1-(4-acetylsulfonyl)phenyltriazene (ASPT-2)

(iii) 3-hydroxy-3-(3-methylphenyl)-1-(4-acetylsulfonyl)phenyltriazene (ASPT-3)

(iv) 3-hydroxy-3-(4-methylphaneyl)-1-(4-acetylsulfonyl)phenyltriazene (ASPT-4)

(v) 3-hydroxy-3-(3-chlorophenyl)-1-(4-acetylsulfonyl)phenyltriazene (ASPT-5)

(vi) 3-hydroxy-3-(4-chlorophenyl)-1-(4-acetylsulfonyl)phenyltriazene (ASPT-6)

(vii) 3-hydroxy-3-(4-carboxyphenyl)-1-(4-acetylsulfonyl)phenyltriazene (ASPT-7)

(viii) 3-hydroxy-3-methyl-1-(4-acetylsulfonyl)phenyltriazene (ASPT-8)

(ix) 3-hydroxy-3-ethyl-1-(4-acetylsulfonyl)phenyltriazene (ASPT-9)

(x) 3-hydroxy-3-(1-methylethyl)-1-(4-acetylsulfonyl)phenyltriazene (ASPT-10)

(xi) 3-hydroxy-3-propyl-1-(4-acetylsulfonyl)phenyltriazene (ASPT-11) and some other 3-alkyl or aryltriazenes having para methoxy and ethoxy substituents in the phenyl group at the –1 position has been reported, and all these compounds have been reported to possess bacteriostatic activity. Further, V(V) complexes of these hydroxytriazenes were reported to show greater antitubercular properties compared to their ligands. Although the analytical application of hydroxytriazenes has been fully explored, medicinal applications as well as biological activities are of recent origin. The focus on their potential as medicinal compounds has been further enhanced by excellent theoretical approaches for predicting a host of activities through ADMET, PASS online, and docking, etc. Research groups have also explored the computer-aided drug designing models which have shown reasonably good results for the validation of predicted activities. Although the number of activities predicted is very large and it is difficult to screen all of them, we have been trying to validate experimentally highly probable activities.

In the last section, a sample of PASS predictions for 3-hydroxy-1,3-diphenyltriazene (HDPT) was included in Table 3.1 to show the scope of prediction of activities for a parent compound. As a general conclusion, it is evidently clear that the compounds of this series possess anti-inflammatory, antioxidant, and antibacterial or antifungal activities at good levels. Not only this, but in our earlier studies and work published by other authors, insecticidal and acaricidal activities have been shown at excellent levels. Thus, this series of compounds has been surprising both medicinal as well as biochemists for their potential as drug candidates which needs further exhaustive exploration. Although similar compounds of the hydroxytriazene series are excellent bioactive candidates, 3-hydroxy-1,3-diphenyltriazene and its analogues are underexplored, and therefore this chapter will offer a state of the art of the compounds in terms of their potential as medicinal compounds.

The earliest report on the acaricidal and insecticidal properties of about 40 alkyl and aryl hydroxytriazenes appeared in 1970 [30]. A U.S. patent on 3-3, dialkyl-1-(substituted-phenyl) triazne-1-oxides along with N-phenylazoheterocycle compounds reports anti-inflammatory properties of hydroxytriazenes [31]. The author and his research group have been working on the biological activities and medicinal applications of the compounds for the last two decades. The synthesis and antimicrobial activities of 13 hydroxytriazenes were reported in 2001 by Goswami and Purohit [32]. The compounds have been reported to possess excellent activities, both antibacterial as well as antifungal. The authors have explained the mechanism of activities on the basis of the chelating properties of the compounds, through blocking of the cell wall by forming chelates they hamper cell-wall synthesis and thus act as antibacterials or antifungals. It has also been highlighted that carboxy and methyl groups attached to 1 or 3 side nitrogens play a major role in enhancing the activities per se.

$$R-\underset{|}{N}-OH$$
$$N=N-R'$$

Figure 3.2 R = alkyl/aryl group and R' = phenyl group.

However, this was the first organized study exploring the possibility of their medicinal potential (Figure 3.2).

Although biological activities including insecticidal activities have been reported in subsequent years, major screening in terms of the medicinal application of hydroxytriazenes has gained interest in the last couple of years only. As evidence, the synthesis and insecticidal activity of some hydroxytriazenes and their vanadium complexes were reported by Goswami [33], explaining that metal complexes could also be used as effective bioactive agents. Although complexes have been shown to exhibit less activity compared to ligands, they have been active. The mechanism discussed also explains with limitation the possibility of the entire vanadium complex to chelate and exert biological activity. The available literature though explains the contrary, suggesting that the biological effect of any synthetic chelate indicates that whenever the chelating ligand is saturated by coordinating with a metal it does not exert much effect in any biological system. This is due to the reduction of the positive charge at the metal atom by the withdrawal of electrons from the ligand. However, the author asserts that when a complex acts it must bind with the cell-wall ingredients, as such blocking the cell-wall synthesis. It opens a new area of research on medicinal applications of hydroxytriazene metal complexes.

The antibacterial and antifungal activity of hydroxytriazenes has been reported by Goswami et al. [24]. As reported, the compounds have shown moderate to excellent activities, and antibacterial activity is better than antifungal, as indicated by minimum inhibitory concentration (MIC) values. To consolidate the work reported on hydroxytriazenes by our group and elsewhere, Figure 3.3 gives a brief account under one heading.

Chauhan et al. [35] have reported the inflammatory activity of some substituted hydroxytriazenes on Wister rats. It is reported that though parent compound 3-hydroxy-1,3-diphenyltriazene did not show any significant activity after 1 h, other substituted compounds, viz. 3-hydroxy-3(m-chloro)phenyl-1-(4-sulfonamide) phenyltriazene and 3-hydroxy-3-(p-chloro)phenyl-1-(4-sulfonamide) phenyltriazene, showed significant anti-inflammatory activity. The maximum inhibition of edema at 1 hour was shown by all the compounds, yet the parent compound was effective only up to 1 h, whereas the other compounds showed inhibition up to 5 h. It is reported that hydroxytriazenes possess significant anti-inflammatory effects on both acute and sub-acute inflammation.

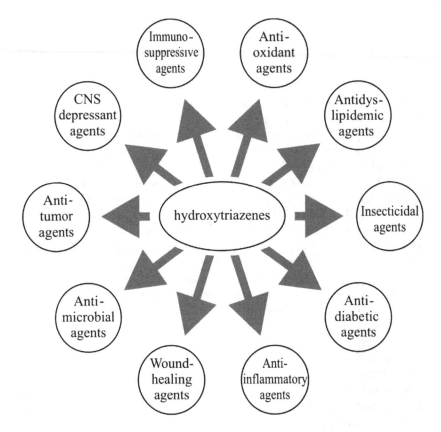

Figure 3.3 Summary of medicinal applications of hydroxytriazenes in the recent past.

Chauhan et al. [36] have reported wound-healing activity of some substituted hydroxytriazenes. The authors used excision, resutured incision, and dead space models. It has been described by the authors that the healing of the wound was assessed by the rate of wound contraction, epithelialization time, skin-breaking strength, granulation strength, etc. It is mentioned that the compounds significantly promote the wound-healing process.

Chauhan et al. [37] have reported the synthesis and analgesic activity of hydroxytriazenes. The study was reportedly performed by the tail immersion method and the acetic acid-induced writhing test. It has been reported that substituted hydroxytriazenes possess significant analgesic activity as revealed by both experimental models.

Singh et al. [38] have reported the anti-inflammatory activity of hydroxytriazenes and their vanadium complexes. The authors reported that both hydroxytriazenes and their vanadium complexes exhibit significant anti-inflammatory activity using the carrageenan-induced hind paw edema method in albino rats. The results report the maximum inhibition of paw volume up to 1 h which reduces with time.

Kumar et al. [39] have reported studies on the insecticidal activity of nine hydroxytriazenes against *Drosophila melanogaster* Meig (fruit fly). It has been reported that the compounds possess good to moderate activity ranging from LC_{50} 0.987 to 5.52 ppm. The residual film method was used, and the probable mechanism has been reported as a contact poison.

Naulakha et al. [40] have reported the synthesis and insecticidal activity of some substituted hydroxytriazenes against *Drosophila melangaster* Meig LC_{50} values reported vary from 9.53 to 56.88 ppm and the mechanistic aspect has been presented. It is mentioned that the compounds act as contact poisons.

Naulakha et al. [41] have reported antifungal activity of hydroxytriazenes and their Cu(II) complexes against *Rhizoctonea solani*, and the percentage inhibition of mycelial growth has been reported as 84.4 to 45.8 which is excellent to moderate. The results have been interpreted in terms of the theory of drug action proposed by Paton.

Using excision, resutured incision, and dead space wound models, Chauhan et al. [42] have reported the wound-healing activity of hydroxytriazenes as a new class of bioactive compounds. The healing of the wound as described by the authors was assessed by the rate of wound concentration, epithelialization, skin breaking strength, granulation strength, dry granulation tissue weight, and finally estimation of hydroxyproline. It has been described that out of three compounds two of them show promotion of wound-healing activity to significant levels.

The synthesis and analgesic activity of some substituted hydroxytriazenes has been reported by Chauhan et al. [43]. It has been described by the authors that the study was performed using the tail-immersion method and the acetic acid induced writhing test. It is described further that although the parent compound 1,3-diphenyl-hydroxytriazene (HDPT) does not show analgesic activity, the other substituted compounds exhibit significant analgesic activity in both experimental methods used.

Joshi et al. [44] have reported polarographic determination and antifungal activity of the Cu(II) complex with 3-hydroxy-3-m-tolyl-1-p-(sulfonamide) phenyltriazene. An attempt is made to correlate electrochemical behavior with activity. A further ligand as well as its Cu(II) complex has been screened against *Rhizoctonic solani*, a fungi causing disease in fennel. The results have been interpreted in terms of percent inhibition (PI) of mycelial growth and sclerotium formation at 500 ppm. It is stated that the complex with Cu(II) has a better PI value (87.03 ppm) compared to the ligand (79.63 ppm). The mechanism of action has been interpreted in terms of the theory of drug action proposed by Paton.

Chauhan et al. [45] have reported the synthesis and antibacterial and antifungal activity of six substituted hydroxytriazenes. It has been mentioned that the compounds have shown very encouraging results, showing a novel application of emerging bioactive compounds. Four bacterial and two fungal strains were used for the screening. The strains used are *E. coli, Salmonella typhi, Pseudomonas aeruginosa, Bacillus subtilis, Aspergilus fumigatus,* and *Candida albicans.*

Gokaraju et al. [46] have reported the preparation and application of triazene analogs in the treatment of metastatic malignant melanoma and other cancers.

Domingues, Vaness O. et al. [47] have reported DNA cleavage, antibacterial activity, and cytotoxicity of triazenes against acute myeloid leukemia cells. Further, it is mentioned that the compounds show cytotoxic activity against myeloid leukemia cells, while some of these exhibit high activity against *B. cerus.*

Osmak, Maja et al. [48] have reported an analog of 1,3-bis (4-nitrophenyl) triazenes, their pharmaceutically acceptable salts, and N-acyl derivatives for tumor treatment. It is mentioned that the compounds can be useful for the treatment of tumor patients both as a single drug as well as in combination with other cytostatics.

A new class of antitumor triazeneoazaindoles has been reported by Diana Patrizia et al. [49]. It is reported that a new class bearing the triazenoazaindole moiety was synthesized for the purpose of identifying antiproliferative agents. The compounds were screened against a panel of human tumor cell lines, and two compounds have been reported to show cytotoxicity in all cell lines, including maintaining this in some multi drug-resistant cell lines. It is briefly explained by flow cytometry analysis that cell-death induction is by apoptosis involving lysosomes.

Toxicity studies of substituted hydroxytriazenes on rodents have been reported by Chauhan et al. [50]. The authors report acute and sub-acute toxicity in rodents. LD_{50} and behavioral activity have been investigated. The study revealed that hydroxytriazenes have no significant influence on the hematological parameters although some biochemical parameters are changed significantly.

Ombaka et al. [51] have reported the synthesis and insecticidal activity of some selected hydroxytriazenes. The activity against a 1-day-old male *Drosophila melanogaster* Meig (fruit fly) has been reported with LC_{50} values ranging from 1.812 to 2.898 ppm, against standard of a commercial product Heptachlor (LC_{50} 1.570 ppm).

Unsalan et al. [52] have reported the synthesis and characterization along with the cytotoxic and antitumor activities of novel triazenes derived from sulfonamides. The effects on A549 and L-929 cell lines were investigated by the authors and no activity was observed.

Cimbora et al. [53] have reported the biological evaluation and anti-tumor activity of 4-nitro substituted 1,3-diaryltriazenes as a potent class synthesis of 1,3-diaryltriazenes and SAR studies have been done, and it is mentioned that this series can be modified from inactive to highly cytotoxic compounds by the introduction of two nitro groups at the para-positions of benzene rings and two additional electron withdraw-ing groups such as bromo, chloro, trifloromethyl, or fluoro substituents at their ortho position. Various acyl groups if added to triazenenitrogen further increase the stability of modified compounds. The authors state that N-acyltriazenes can be considered as prodrugs of non-acylated tri-azenes. The major highlight of the investigation is their high cytotoxicity against different tumor cell lines including cis-platin-resistant laryngeal carcinoma cells. DNA binding analysis has been done and suggests minor groove binding. It is suggested by the authors that 1,3-diaryltriazenes are a new class of anticancer molecules which preferentially target malignant cells and can be potent antitumor drugs.

The synthesis and neuropharmacological effects of some substituted hydroxytriazenes have been reported by Chauhan et al. [54]. The authors report some neuropharmacological effects of synthesized hydroxytri-azenes which include behavior, locomotion, sleep, and convulsions using standard methods on mice. It is further described that the compounds exhibited reduced alertness, reduced spontaneous locomotion activity, and potentiated the pentabarbitone induced sleep. Thus it has been con-cluded that hydroxytriazenes show CNS-depressant effects.

Khanum et al. [55] have reported the synthesis, characterization, and antifungal activity of 3-hydroxy-3-p-tolyl-1-m-nnitrophenyltriazenes and its iron(III) complex. It is reported that the compound has inhibitory action against three fungi, viz., *Rhizoctonia solani, Fusarium solani,* and *Pythium.* Prabhat KB et al. [56] have reported the synthesis, characterization, and activity prediction of a new class of hydroxytriazenes. The series has been studied originating from 4-aminoantipyrine-based hydroxytriazenes. Only theoretical predictions using PASS, synthesis, and characterization of this class of hydroxytriazenes have been reported.

Babel et al. [57] have reported the synthesis, activity prediction, and spectrophotometric study of molybdenum complex of 3-hydroxy-3-p-tolyl-1-p-carboxyphenyltriazenes. Apart from the spectrophotometric studies of molybdenum complexes, theoretical prediction using PASS software has been described showing values ranging from Pa $\sim$ 0.881 to Pa $\sim$ 0.741.

Bhandari et al. [58] have reported the antimicrobial evaluation of some hydroxytriazenes and their ternary metal complexes. An attempt has been made to screen both ligands, that is, hydroxytriazenes and their ternary complexes, with salicylaldehyde using Cu(II) as the metal. The interesting study has explored not only metal complexes as potent anti-bacterial/antifungal agents but also ternary complexes of this series for the first time. Ciprofloxacin and flucanozole have been used as standards. The zone inhibition method was reportedly used.

Singh et al. [59] have reported the biological evaluation of some hydroxytriazenes. Using PASS probable activities of four hydroxytriazenes, viz. 3-hydroxy-3-phenyl-1p-chlorophenyltriazene, 3-hydroxy-3-p-tolyl-1-phenyltriazene, 3-hydroxy-3-p-tolyl-1-p-1-p-chlorophenyltriazene, and 3-hydroxy-3-p-chloro-1-p-nitrophenyltriazenes, have been screened for antibacterial activity out of predicted ones. Using the cup and well method the zones of inhibition have been reported for *E. coli, P. aerugithosa, P. mirabilis,* and *B. subtilis,* and the activities have been found to be low to moderate. However, it is concluded by the authors that PASS may estimate the pharmacotherapeutic potential, possible molecular mechanisms of action, toxic effects, etc., for the compounds a priori.

Baroliya et al. [60] have reported the synthesis, characterization, and antimicrobial activities of hydroxytriazenes and their Co(II) complexes. The paper describes the synthesis of fluoro and chloro substituted hydroxytriazenes and their Co(II) complexes. The ligands as well as their Co(II) complexes have been screened against four bacterial and six fungal strains. The authors describe that both the ligand as well as Co(II) complex show good to moderate antimicrobial activity (< 50 ppm).

Khanum et al. [61] have reported the antifungal activity of hydroxytriazenes and their ternary complexes with vanadium (V) and thiourea. The authors have reported the synthesis of 3-hydroxy-3-m-chlorophenyl-1-(4-sulfohamidophenyl)triazene, 3-hydroxy-3-n-propyl-1-(4-sulfnamidophenyltriazene, 3-hydroxy-3-isopropyl-1-(4-sulfoamidophenyl)triazene, and 3-hydroxy-3-m-tolyl-1-p-chlorophenyltriazene and screened these compounds against two fungi, *Candida albicans* and *Aspergillus fumigatus.* The inhibitory activity has been reported at 100, 200, 500, and 1000 ppm levels. This is a novel application of this emerging class of bioactive molecules as reported by the authors.

Khanum et al. [62] have reported the synthesis and insecticidal activity of some hydroxytriazene derivatives and their ternary complexes with thiourea of vanadium (V). The compounds as reported have been screened for their insecticidal activity against adult *Chrotogonus trachypterus* Blanch (surface grasshopper).

Chundawat et al. [63] have reported the antibacterial activity of hydroxytriazenes, Schiff's bases and their ternary complexes of Zn(II). The authors report activity against *E. coli, P. mirabilus, S. aureus,* and *S. facealis* using the agar well method. It has been reported that the mixed ligand complexes have relatively enhanced activity compared to their corresponding ligands.

A very interesting report on the synthesis, characterization, theoretical prediction, and evaluation of biological activities of some sulfacetamide-based hydroxytriazenes has been published by Shilpa Agarwal et al. [64].

The main objective of the paper is to establish a computer-aided drug design protocol for hydroxytriazenes. Six new sulfacetamide-based hydroxytriazenes have been synthesized, and their theoretical prediction using PASS has been done. Out of a number of theoretically predicted activities the authors have chosen to validate experimentally the anti-inflammatory, anti-radical, and antidiabetic activities of the six new compounds. The author have further mentioned that the compounds have shown moderate to excellent anti-inflammatory, anti-radical, and antidiabetic activities, and concluded that hydroxytriazenes can prove excellent potential candidates for future drugs and particularly multi-targeted ones. The study also helps in developing a synthetic protocol for theoretically predicted activities of hydroxytriazenes.

A very interesting study on the antidyslipidemic and antioxidant effects of novel hydroxytriazenes has been published by Regar et al. [65]. The lipid-lowering effect of synthesized compounds has been evaluated on a triton-induced hyperlipidemia model in adult male Charles Foster rats. The authors report that the administration of 400 mg/kg body weight dose of triton WR 1339 causes an increase in total cholesterol, phospholipid, and triglyceride levels, inhibiting post-heparin lipolytic activity, which was reversed by hydroxytriazenes. The dose of 100 mg/kg body wt also inhibits the generation of free radicals, viz. superoxide anions and hydroxyl radicals. The lipid oxidation has also been reported to be inhibited.

A condensed review on the anti-inflammatory activity has been published by Goswami et al. [66]. The review is comprised of work done on 18 different hydroxytriazenes possessing anti-inflammatory activity. The review highlights hydroxytriazenes as potential anti-inflammatory agents which can act as multi-targeted drugs in the future.

3.5 Summary of medicinal applications and biological activities of hydroxytriazenes

Hydroxytriazenes, having been exhaustively used as analytical reagents, as evidenced by voluminous reviews both as spectrophotometric reagents and complexometric/metallochromic indicators, finally have been surprisingly showing promising medicinal and biological applications. Starting from antimicrobial to cytotoxic activities, this interesting chemical moiety has exhibited enormous potential as a drug candidate. The work done by our group and elsewhere signified the need to consolidate the approach of research on this series of compounds. However, out of various activities shown by hydroxytriazenes the most promising activities in recent years are their anti-inflammatory as well as anti-radical or antidiabetic ones. There is no doubt that a computer-aided strategy to further find out some really good

Table 3.3 Medicinal/biological applications of hydroxytriazenes

S. No.	IUPAC name	Structure	Activity	Reference
1	3-Hydroxy-3-p-tolyl-1-o-carboxyphenyltriazene	(structure)	Antibacterial Antifungal	Goswami et al. (2001)
2	3-Hydroxy-3-o-tolyl-1-o-carboxyphenyltriazene	(structure)	Antibacterial Antifungal	Goswami et al. (2001)
3	3-Hydroxy-3-m-tolyl-1-o-carboxyphenyltriazene	(structure)	Antibacterial Antifungal	Goswami et al. (2001)
4	3-Hydroxy-3-p-tolyl-1-p-chlorophenyltriazene	(structure)	Antibacterial Antifungal Insecticidal	Goswami et al. (2001) Goswami et al. (2002)
5	3-Hydroxy-3-m-tolyl-1-p-chlorophenyltriazene	(structure)	Antibacterial Antifungal Insecticidal	Goswami et al. (2001) Ombaka AO et al. (2011)
6	3-Hydroxy-3-n-propyl-1-p-chlorophenyltriazene	(structure)	Antibacterial Antifungal	Goswami et al. (2001)

(*Continued*)

Table 3.3 (Continued) Medicinal/biological applications of hydroxytriazenes

S. No.	IUPAC name	Structure	Activity	Reference
7	3-Hydroxy-3-p-tolyl-1-o-chlorophenyltriazene		Antibacterial Antifungal	Goswami et al. (2001)
8	3-Hydroxy-3-m-tolyl-1-o-chloro-phenyltriazene		Antibacterial Antifungal Insecticidal	Goswami et al. (2001) Ombaka AO et al. (2011)
9	3-Hydroxy-3-m-tolyl-1-p-nitrophenyltriazene		Antibacterial Antifungal	Goswami et al. (2001)
10	3-Hydroxy-3-m-tolyl-1-p-tolyltriazene		Antibacterial Antifungal	Goswami et al. (2001)
11	3-Hydroxy-3-m-tolyl-1-m-hydroxyphenyltriazene		Antibacterial Antifungal	Goswami et al. (2001)

(Continued)

Table 3.3 (Continued) Medicinal/biological applications of hydroxytriazenes

S. No.	IUPAC name	Structure	Activity	Reference
12	3-Hydroxy-3-methyl-1-(2,4,6-tribromophenyl)triazene		Antibacterial Antifungal	Goswami et al. (2001)
13	3-Hydroxy-3-m-tolyl-1-phenyltriazene		Antibacterial Antifungal Insecticidal	Goswami et al. (2001) Bhandari A et al. (2012) Ombaka AO et al. (2011)
14	3-Hydroxy-3-phenyl-1-p-carboxyphenyltriazene		Insecticidal	Goswami et al. (2002)
15	3-Hydroxy-3-phenyl-1-p-tolyltriazene		Insecticidal	Goswami et al. (2002)
16	3-Hydroxy-3-phenyl-1-m-methoxyphenyltriazene		Insecticidal	Goswami et al. (2002)
17	3-Hydroxy-3-phenyl-1-m-acetaphenontriazene		Insecticidal	Goswami et al. (2002)

(Continued)

Table 3.3 (Continued) Medicinal/biological applications of hydroxytriazenes

S. No.	IUPAC name	Structure	Activity	Reference
18	3-Hydroxy-1,3-diphenyltriazene		Anti-inflammatory Analgesic Anti-inflammatory Antifungal activity Antibacterial Antimicrobial activity	Chahan LS et al. (2006) Chauhan LS et al. (2007) Singh K et al. (2008) Naulakha N et al. (2009) Chahan LS et al. (2010) Singh GP et al. (2013)
19	3-Hydroxy-3-phenyl-1-(4-sulfonamide)phenyltriazene		Anti-inflammatory Analgesic Anti-inflammatory Insecticidal Antifungal activity Wound healing Antibacterial	Chahan LS et al. (2006) Chauhan LS et al. (2007) Singh K et al. (2008) Naulakha N et al. (2008) Naulakha N et al. (2008) Chauhan LS et al. (2006) Chauhan LS et al. (2010)
20	3-Hydroxy-3-(m-chloro)phenyl-1-(4-sulfonamide)phenyltriazene		Anti-inflammatory Analgesic Anti-inflammatory Antifungal Antibacterial	Chauhan LS et al. (2006) Chauhan LS et al. (2007) Singh K et al. (2008) Chauhan LS et al. (2010)
21	3-Hydroxy-3-(p-chloro)phenyl-1-(4-sulfonamide)phenyltriazene		Anti-inflammatory Analgesic Anti-inflammatory Antifungal Antibacterial	Chauhan LS et al. (2006) Chauhan LS et al. (2007) Singh K et al. (2008) Chauhan LS et al. (2010)

(Continued)

Table 3.3 (Continued) Medicinal/biological applications of hydroxytriazenes

S. No.	IUPAC name	Structure	Activity	Reference
22	3-Hydroxy-3-propyl-1-(4-sulfonamide) phenyltriazene	H_2NO_2S—⬡—N=N—N—C_3H_7 (OH)	Insecticidal Wound healing Antifungal Antibacterial Antifungal Antibacterial	Naulakha N et al. (2008) Chauhan LS et al. (2010) Chauhan LS et al. (2010) Sharma D et al. (2012)
23	3-Hydroxy-3-isopropyl-1-(4-sulfonamide) phenyltriazene	H_2NO_2S—⬡—N=N—N—CH(CH_3)(CH_3) (OH)	Insecticidal Wound healing Antifungal Antibacterial Antifungal Antibacterial	Naulakha N et al. (2008) Chauhan LS et al. (2010) Chauhan LS et al. (2010) Sharma D et al. (2012)
24	3-Hydroxy-3-o-tolyl-1-(4-sulfonamide) phenyltriazene	H_2NO_2S—⬡—N=N—N—⬡(CH_3) (OH)	Insecticidal	Naulakha N et al. (2008)
25	3-Hydroxy-3-m-tolyl-1-(4-sulfonamide) phenyltriazene	H_2NO_2S—⬡—N=N—N—⬡(CH_3) (OH)	Insecticidal Antifungal	Naulakha N et al. (2008) Joshi P et al. (2010)
26	3-Hydroxy-3-p-tolyl-1-(4-sulfonamide) phenyltriazene	H_2NO_2S—⬡—N=N—N—⬡—CH_3 (OH)	Insecticidal	Naulakha N et al. (2008)

(Continued)

Table 3.3 (Continued) Medicinal/biological applications of hydroxytriazenes

S. No.	IUPAC name	Structure	Activity	Reference
27	3-Hydroxy-3-o-sulfonato (sodium salt)phenyl-1-(4-sulfonamide) phenyltriazene		Insecticidal	Naulakha N et al. (2008)
28	3-Hydroxy-3-p-tolyl-1-phenyltriazene		Insecticidal Antibacterial Antifungal	Naulakha N et al. (2008) Bhandari A et al. (2012)
29	3-Hydroxy-3-p-tolyl-1-o-nitrophenyltriazene		Insecticidal	Naulakha N et al. (2008)
30	3-Hydroxy-3-m-sulfonato (sodium salt) phenyl-1-o-nitrophenyltriazene		Insecticidal	Kumar S et al. (2009)
31	3-Hydroxy-3-phenyl-1-o-chlorophenyltriazene		Insecticidal Antibacterial Antifungal	Kumar S et al. (2009) Bhandari A et al. (2012)

(Continued)

Table 3.3 (Continued) Medicinal/biological applications of hydroxytriazenes

S. No.	IUPAC name	Structure	Activity	Reference
32	3- Hydroxy-3-m-sulfonato (sodium salt)-1-o-chlorophenyltriazene		Insecticidal	Kumar S et al. (2009)
33	3-Hydroxy-3-n-propyl-1-p-methylphenyltriazene		Insecticidal	Kumar S et al. (2009)
34	3- Hydroxy-3-ethyl-1-1-(4- sulponamidophenyl) triazene		Insecticidal	Kumar S et al. (2009)
35	3- Hydroxy-3-methyl-1-m-nitrophenyltriazene		Insecticidal	Kumar S et al. (2009)
36	3- Hydroxy-3-p-tolyl-1-p-nitrophenyltriazene		Insecticidal Antifungal	Kumar S et al. (2009) Naulakha N et al. (2009)
37	3-Hydroxy-3-n-proply-1-m-chlorophenytrianzene		Insecticidal Antimicrobial Antimicrobial	Kumar S et al. (2009) Ombaka AO et al. (2012) Singh GP et al. (2013)

(Continued)

Table 3.3 (Continued) Medicinal/biological applications of hydroxytriazenes

S. No.	IUPAC name	Structure	Activity	Reference
38	3-Hydroxy-3-p-tolyl-1-p-sulfonato(sodium salt) phenyltriazene		Insecticidal	Kumar S et al. (2009)
39	3-Hydroxy-3-p-tolyl-1-m-nitrophenyltriazene		Antifungal	Khanam R et al. (2011)
40	3-Hydroxy-3-m-tolyl-1-m-nitrophenyltriazene		Insecticidal	Ombaka AO et al. (2011)
41	3-Hydroxy-3-m-tolyl-1-p-methoxyphenyltriazene		Insecticidal	Ombaka AO et al. (2011)
42	3-Hydroxy-3-phenyl-1-p-chlorophenyltriazene		Insecticidal	Ombaka AO et al. (2011)
43	3-Hydroxy-3-n-propyl-1-o-chlorophenyltriazene		Insecticidal	Ombaka AO et al. (2011)

(Continued)

Table 3.3 (Continued) Medicinal/biological applications of hydroxytriazenes

S. No.	IUPAC name	Structure	Activity	Reference
44	3-Hydroxy-3-m-tolyl-1-o-carboxyxyphenyltriazene		Insecticidal	Ombaka AO et al. (2011)
45	3-Hydroxy-3-phenyl-1-pchlorophenytrianzene		Antibacterial Antifungal	Bhandari A et al. (2012)
46	3-Hydroxy-3-phenyl-1-3-hydroxyphenyl)triazene		Antibacterial Antifungal	Sharma D et al. (2012)
47	3-Hydroxy-3-phenyl-1-5-chloro-2-hydroxy-phenyl) triazene		Antibacterial Antifungal	Sharma D et al. (2012)
48	3-Hydroxy-3-phenyl-1-(2,6-dichloro-4-trifluoro-methylphenyl) triazene		Antibacterial Antifungal	Sharma D et al. (2012)

(Continued)

Table 3.3 (Continued) Medicinal/biological applications of hydroxytriazenes

S. No.	IUPAC name	Structure	Activity	Reference
49	3-Hydroxy-3-p-tolyl-1-(2,4,6-tribromophenyl)triazene		Antibacterial Antifungal	Ombaka AO et al. (2012)
50	3-Hydroxy-3-m-tolyl-1-(2,4,6-tribromophenyl)triazene		Antibacterial Antifungal	Ombaka AO et al. (2012)
51	3-Hydroxy-3-m-tolyl-1-(3,4-dichlorophenyl)triazene		Antibacterial Antifungal	Ombaka AO et al. (2012)
52	3-Hydroxy-3-o-tolyl-1-p-chlorophenyltriazene		Antibacterial Antifungal	Ombaka AO et al. (2012)
53	3-Hydroxy-3-o-tolyl-1-o-chlorophenyltriazene		Antibacterial Antifungal Insecticidal	Ombaka AO et al. (2012) Rezaie B et al. (1997)

(Continued)

Table 3.3 (Continued) Medicinal/biological applications of hydroxytriazenes

S. No.	IUPAC name	Structure	Activity	Reference
54	3-Hydroxy-3-n-propyl-1-m-chlorophenyltriazene		Antibacterial Antifungal	Ombaka AO et al. (2012)
55	3-Hydroxy-3-p-tolyl-1-m-hydroxyphenyltriazene		Antibacterial Antifungal	Ombaka AO et al. (2012)
56	3-Hydroxy-3-m-tolyl-1-o-nitrophenyltriazene		Antibacterial Antifungal	Ombaka AO et al. (2012)
57	3-Hydroxy-3-p-tolyl-1-o-carboxyphenyltriazene		Antibacterial Antifungal	Ombaka AO et al. (2012)
58	3-Hydroxy-3-o-tolyl-1-o-carboxyphenyltriazene		Antibacterial Antifungal	Ombaka AO et al. (2012)
59	3-Hydroxy-3-phenyl-1-p-sulfonato(sodium salt)phenyltriazene		Antimicrobial	Singh GP et al. (2013)

(Continued)

Table 3.3 (Continued) Medicinal/biological applications of hydroxytriazenes

S. No.	IUPAC name	Structure	Activity	Reference
60	3-Hydroxy-3-o-tolyl-1-p-sulfonato(sodium salt) phenyltriazene		Antimicrobial	Singh GP et al. (2013)
61	3-Hydroxy-3-m-tolyl-1-p-sulfonato(sodium salt) phenyltriazene		Antimicrobial	Singh GP et al. (2013)
62	3-Hydroxy-3-p-aceto phenyl-1-p-sulfonato (sodium salt) henyltriazene		Antimicrobial	Singh GP et al. (2013)
63	3-Hydroxy-3-p-tolyl-1-p-acetanilidephenyltriazene		Antimicrobial	Singh GP et al. (2013)
64	3-hydroxy-3-phenly-1-(2,5-dichlorophenyl) triazene		Antimicrobial	Kodli KK et al. (2014)
65	3-Hydroxy-3-m-tolyl-1-(2,5 dichlorophenyl) triazene		Antimicrobial	Kodli KK et al. (2014)

(Continued)

Table 3.3 (Continued) Medicinal/biological applications of hydroxytriazenes

S. No.	IUPAC name	Structure	Activity	Reference
66	3-Hydroxy-3-p-tolyl-1-(2,4 dichlorophenyl) triazene		Antimicrobial	Kodli KK et al. (2014)
67	3-Hydroxy-3-isopropyl-1-(2,4-dichlorophenyl) triazene		Antimicrobial	Kodli KK et al. (2014)
68	3-Hydroxy-3-phenyl-1-(3-chloro-4-fluorophenyl) triazene		Antimicrobial	Baroliya PK et al. (2014)
69	3-Hydroxy-3-(2-methyl phenyl)-1-(3-chloro-4-fluorophenyl) triazene		Antimicrobial	Baroliya PK et al. (2014)
70	3-Hydroxy-3-(3-methyl-phenyl)-1-(3-chloro-4-fluorophenyl) triazene		Antimicrobial	Baroliya PK et al. (2014)
71	3-Hydroxy-3-(4-methyl phenyl)-1-(3-chloro-4-fluorophenyl) triazene		Antimicrobial	Baroliya PK et al. (2014)

(Continued)

Table 3.3 (Continued) Medicinal/biological applications of hydroxytriazenes

S. No.	IUPAC name	Structure	Activity	Reference
72	3-Hydroxy-3-methyl-1-[N'-(1-phenylethylidene) pyridine-3-carbohydrazide] triazene		Anti-inflammatory	Patidar AK et al. (2011)
73	3-Hydroxy-3-ethyl-1-[N'-(1-phenylethylidene) pyridine-3-carbohydrazide] triazene		Anti-inflammatory	Patidar AK et al. (2011)
74	3-Hydroxy-3-n-propyl-1-[N'-(1-phenylethylidene) pyridine-3-carbohydrazide] triazene		Anti-inflammatory	Patidar AK et al. (2011)
75	3-Hydroxy-3-isopropyl-1-[N'-(1-phenylethylidene) pyridine-3-carbohydrazide] triazene		Anti-inflammatory	Patidar AK et al. (2011)
76	3-Hydroxy-3-methyl-1-[N'-(1-phenylethylidene) pyridine-4-carbohydrazide] triazene		Anti-inflammatory	Patidar AK et al. (2011)

(Continued)

Table 3.3 (Continued) Medicinal/biological applications of hydroxytriazenes

S. No.	IUPAC name	Structure	Activity	Reference
77	3-Hydroxy-3-ethyl-1-[N′-(1-phenylethylidene) pyridine-4-carbohydrazide] triazene		Anti-inflammatory	Patidar AK et al. (2011)
78	3-Hydroxy-3-n-propyl-1-[N′-(1-phenylethylidene) pyridine-4-carbohy-drazide] triazene		Anti-inflammatory	Patidar AK et al. (2011)
79	3-Hydroxy-3-isopropyl-1-[N′-(1-phenylethylidene) pyridine-4-carbohydrazide] triazene		Anti-inflammatory	Patidar AK et al. (2011)
80	3-Hydroxy-3-phenyl-1-(4-acetylsulfonyl) phenyltriazene		Antioxidant Antidiabetic	Agrawal S et al. (2016)
81	3-Hydroxy-3-(3-methylphenyl)-1-(4-acetylsulfonyl) phenyltriazene		Antioxidant Antidiabetic	Agrawal S et al. (2016)
82	3-Hydroxy-3-(4-chlorophenyl)-1-(4-acetylsulfonyl) phenyltriazene		Antioxidant Antidiabetic	Agrawal S et al. (2016)

(Continued)

Table 3.3 (Continued) Medicinal/biological applications of hydroxytriazenes

S. No.	IUPAC name	Structure	Activity	Reference
83	3-Hydroxy-3-methyl-1-(4-acetylsulfonyl)phenyltriazene		Antioxidant Antidiabetic	Agrawal S et al. (2016)
84	3-Hydroxy-3-ethyl-1-(4-acetylsulfonyl)phenyltriazene		Antioxidant Antidiabetic	Agrawal S et al. (2016)
85	3-Hydroxy-3-(1-methylethyl)-1-(4-acetylsulfonyl)phenyltriazene		Antioxidant Antidiabetic	Agrawal S et al. (2016)
86	3-Hydroxy-3-phenyl-1-[2-(trifluoromethyl)phenyl]triaz-1-ene		Antioxidant Antidyslipidemic	Regar ML et al. (2016)
87	3-Hydroxy-3-(2-ethylphenyl)-1-[2-(trifluoromethyl)phenyl]triaz-1-ene		Antioxidant Antidyslipidemic	Regar ML et al. (2016)
88	3-Hydroxy-3-(3-ethylphenyl)-1-[2-(trifluoromethyl)phenyl]triaz-1-ene		Antioxidant Antidyslipidemic	Regar ML et al. (2016)

(Continued)

Table 3.3 (Continued) Medicinal/biological applications of hydroxytriazenes

S. No.	IUPAC name	Structure	Activity	Reference
89	3-Hydroxy-3-(4-ethylphenyl)-1-[2-trifluoromethyl)-phenyl]triaz-1-ene		Antioxidant Antidyslipidemic	Regar ML et al. (2016)
90	3-(4-Chlorophenyl)-3-ydroxy-1-[2-(trifluoromethyl)-phenyl] triazene		Antioxidant Antidyslipidemic	Regar ML et al. (2016)
91	3-(4-Acetylphenyl)-3-ydroxy-1-[2-(trifluoromethyl)-phenyl]triaz-1-ene		Antioxidant Antidyslipidemic	Regar ML et al. (2016)
92	3-Hydroxy-3-phenyl-1-[4-(trifluoromethyl) phenyl]triaz-1-ene		Antioxidant Antidyslipidemic	Regar ML et al. (2016)
93	3-Hydroxy-3-(3-ethylphenyl)-1-[4-((trifluoromethyl)-phenyl]triaz-1-ene		Antioxidant Antidyslipidemic	Regar ML et al. (2016)
94	3-Hydroxy-3-(4-ethylphenyl)-1-[4-(trifluoromethyl)-phenyl]triaz-1-ene		Antioxidant Antidyslipidemic	Regar ML et al. (2016)

(Continued)

Table 3.3 (Continued) Medicinal/biological applications of hydroxytriazenes

S. No.	IUPAC name	Structure	Activity	Reference
95	3-Hydroxy-3-methyl-1-(4-sulfonamidophenyl)triazene	H_2N-SO_2-C$_6$H$_4-N=N-N(OH)-CH_3$	Antidiabetic Antioxidant Anti-inflammatory	Not published
96	3-Hydroxy-3-ethyl-1-(4-sulfonamidophenyl)triazene	H_2N-SO_2-C$_6$H$_4-N=N-N(OH)-CH_2-CH_3$	Antidiabetic Antioxidant Anti-inflammatory	Not published
97	3-Hydroxy-3-propyl-1-(4-sulfonamidophenyl)triazene	H_2N-SO_2-C$_6$H$_4-N=N-N(OH)-CH_2-H_2C-CH_3$	Antidiabetic Antioxidant Anti-inflammatory	Not published Not published
98	3-Hydroxy-3-(2-propyl-1-(4-sulfonamidophenyl)triazene	H_2N-SO_2-C$_6$H$_4-N=N-N(OH)-CH(CH_3)CH_3$	Antidiabetic Antioxidant Anti-inflammatory	Not published
99	3-Hydroxy-3-phenyl-1-(4-sulfonamidophenyl)triazene	H_2N-SO_2-C$_6$H$_4-N=N-N(OH)-$C$_6$H$_5$	Antidiabetic Antioxidant Anti-inflammatory	Not published
100	3-Hydroxy-3-(2-methylphenyl)-1-(4-sulfonamidophenyl)triazene	H_2N-SO_2-C$_6$H$_4-N=N-N(OH)-$(2-CH$_3$)C$_6$H$_4$	Antidiabetic Antioxidant Anti-inflammatory	Not published

(Continued)

Table 3.3 (Continued) Medicinal/biological applications of hydroxytriazenes

S. No.	IUPAC name	Structure	Activity	Reference
101	3-Hydroxy-3-(3-methylphenyl)-1-(4-sulfonamidophenyl)triazene		Antidiabetic Antioxidant Anti-inflammatory	Not published
102	3-Hydroxy-3-(4-methylphenyl)-1-(4-sulfonamidophenyl)triazene		Antidiabetic Antioxidant Anti-inflammatory	Not published
103	3-Hydroxy-3-methyl-1-(N-pyrimidin-2-ylbenzene-1-sulfonamide)triazene		Antidiabetic Antioxidant	Not published
104	3-Hydroxy-3-ethyl-1-(N-pyrimidine-2-ylbenzene-1-sulfonamide)triazene		Antidiabetic Antioxidant	Not published
105	3-Hydroxy-3-propyl-1-(N-pyrimidine-2-ylbenzene-1-sulfonamide)triazene		Antidiabetic Antioxidant	Not published
106	3-Hydroxy-3-phenyl-1-(N-pyrimidine-2-ylbenzene-1-sulfonamide)triazene		Antidiabetic Antioxidant	Not published

(Continued)

Table 3.3 (Continued) Medicinal/biological applications of hydroxytriazenes

S. No.	IUPAC name	Structure	Activity	Reference
107	3-Hydroxy-3-(2-methylphenyl)-1-(N-pyrimidine-2-ylbenzene-1-sulfonamide) triazene		Antidiabetic Antioxidant	Not published
108	3-Hydroxy-3-(3-methylphenyl)-1-(N-pyrimidine-2-ylbenzene-1-sulfonamide) triazene		Antidiabetic Antioxidant	Not published
109	3-Hydroxy-3-(4-methylphenyl)-1-(N-pyrimidine-2-ylbenzene-1-sulfonamide) triazene		Antidiabetic Antioxidant	Not published
110	3-Hydroxy-3-propyl-1-(N-pyridine-2-ylbenzene-1-sulfonamide) triazene		Antidiabetic Antioxidant	Not published
111	3-Hydroxy-3-phenyl-1-(N-pyridine-2-ylbenzene-1-sulfonamide) triazene		Antidiabetic Antioxidant Anti-inflammatory	Not published

(Continued)

Table 3.3 (Continued) Medicinal/biological applications of hydroxytriazenes

S. No.	IUPAC name	Structure	Activity	Reference
112	3-Hydroxy-3-(2-methylphenyl)-1-(N-pyridine-2-ylbenzene-1-sulfonamide) triazene		Antidiabetic Antioxidant Anti-inflammatory	Not published
113	3-Hydroxy-3-(3-methylphenyl)-1-(N-pyridine-2-ylbenzene-1-sulfonamide) triazene		Antidiabetic Antioxidant Anti-inflammatory	Not published
114	3-Hydroxy-3-(4-methylphenyl)-1-(N-pyridine-2-ylbenzene-1-sulfonamide) triazene		Antidiabetic Antioxidant Anti-inflammatory	Not published
115	3-Hydroxy-3-ethyl-1-[N-(4,6-dimethylpyrimidin-2-yl)benzene-1-sulfonamid]triazene		Antidiabetic Antioxidant Anti-inflammatory	Not published
116	3-Hydroxy-3-propyl-1-[N-(4,6-dimethylpyrimidin-2-yl)benzene-1-sulfonamid]triazene		Antidiabetic Antioxidant Anti-inflammatory	Not published

(Continued)

Table 3.3 (Continued) Medicinal/biological applications of hydroxytriazenes

S. No.	IUPAC name	Structure	Activity	Reference
117	3-Hydroxy-3-(2-propyl-1-[N-(4,6-dimethylpyrimidin-2-yl)benzene-1-sulfonamid]triazene		Antidiabetic Antioxidant Anti-inflammatory	Not published
118	3-Hydroxy-3-phenyl-1-[N-(4,6-dimethylpyrimidin-2-yl)benzene-1-sulfonamid]triazene		Antidiabetic Antioxidant Anti-inflammatory	Not published
119	3-Hydroxy-3-(3-methylphenyl)-1-[N-(4,6-dimethylpyrimidin-2-yl)-benzene-1-sulfonamid]triazene		Antidiabetic Antioxidant Anti-inflammatory	Not published
120	3-Hydroxy-3-(4-methylphenyl)-1-[N-(4,6-dimethylpyrimidin-2-yl)-benzene-1-sulfonamid]triazene		Antidiabetic Antioxidant Anti-inflammatory	Not published
121	3-Hydroxy-3-ethyl-1-(N-caramimidoylbenzene-1-sulfonamid)triazene		Antidiabetic Antioxidant Anti-inflammatory	Not published

(Continued)

Table 3.3 (Continued) Medicinal/biological applications of hydroxytriazenes

S. No.	IUPAC name	Structure	Activity	Reference
122	3-Hydroxy-3-propyl-1-(N-caramimidoylbenzene-1-sulfonamid) triazene	$H_2N-C-HN-S=O$, $\overset{\parallel}{NH}$, $O=S$, benzene ring, $N=N-N-CH_2$, OH, H_2C-CH_3	Antidiabetic Antioxidant Anti-inflammatory	Not published
123	3-Hydroxy-3-(2-propyl-1-(N-caramimidoylbenzene-1-sulfonamid) triazene	$H_2N-C=HN-S=O$, $\overset{\parallel}{NH}$, $O=S$, benzene ring, $N=N-N-CH$, OH, CH_3, CH_3	Antidiabetic Antioxidant Anti-inflammatory	Not published
124	3-Hydroxy-3-phenyl-1-(N-caramimidoylbenzene-1-sulfonamid) triazene	$H_2N-C-HN-S=O$, $\overset{\parallel}{NH}$, $O=S$, benzene ring, $N=N-N$, OH, phenyl	Antidiabetic Antioxidant Anti-inflammatory	Not published
125	3-Hydroxy-3-(4-methylphenyl)-1-(N-caramimidoylbenzene-1-sulfonamid) triazene	$H_2N-C-HN-S=O$, $\overset{\parallel}{NH}$, $O=S$, benzene ring, $N=N-N$, OH, CH_3	Antidiabetic Antioxidant Anti-inflammatory	Not published

Table 3.4 Medicinal/biological applications of hydroxytriazene/triazene-1-oxide

S. No.	IUPAC name	Structure	Activity	Reference
1	1-Methyl-3-methyl-3-phenyltriazene-1-oxide		Anti-inflammatory	Miesel JL et al. (1976)
2	1-Methyl-3-(3-methoxyphenyl)triazene-1-oxide		Anti-inflammatory	Miesel JL et al. (1976)
3	1-Methyl-3-(3-methyl)triazene-1-oxide		Anti-inflammatory	Miesel JL et al. (1976)
4	1-Methyl-3-(3-pyridyl)triazene-1-oxide		Anti-inflammatory	Miesel JL et al. (1976)
5	1-Methyl-3-phenyltriazene-1-oxide		Anti-inflammatory	Miesel JL et al. (1976)

(Continued)

Table 3.4 (Continued) MeMedicinal/biological application of hydroxytriazene/triazene-1-oxide

S. No.	IUPAC name	Structure	Activity	Reference
6	1-Methyl-3 (m-trifluoromethylphenyl) triazene-1-oxide		Anti-inflammatory	Miesel JL et al. (1976)
7	1,3-Diphenyl triazene-1-oxide		Anti-inflammatory	Miesel JL et al. (1976)
8	1-Phenyl-3-(o-chlorophenyl) triazene-1-oxide		Anti-inflammatory	Miesel JL et al. (1976)
9	1-Phenyl-3-(m-chlorophenyl) triazene-1-oxide		Anti-inflammatory	Miesel JL et al. (1976)
10	1-Phenyl-3-(p-chlorophenyl) triazene-1-oxide		Anti-inflammatory	Miesel JL et al. (1976)

(Continued)

Table 3.4 (Continued) MeMedicinal/biological application of hydroxytriazene/triazene-1-oxide

S. No.	IUPAC name	Structure	Activity	Reference
11	1-Phenyl-3-(2,5-dimethylphenyl) triazene-1-oxide		Anti-inflammatory	Miesel JL et al. (1976)
12	1-Phenyl-3-(m-trifuoromethy-lphenyl) triazene-1-oxide		Anti-inflammatory	Miesel JL et al. (1976)
13	1-Phenyl-3-(m-nitrophenyl) triazene-1-oxide		Anti-inflammatory	Miesel JL et al. (1976)
14	1-Phenyl-3 p-N,N-dimethylphenyl) triazene-1-oxide		Anti-inflammatory	Miesel JL et al. (1976)

(Continued)

Table 3.4 (Continued) MeMedicinal/biological application of hydroxytriazene/triazene-1-oxide

S. No.	IUPAC name	Structure	Activity	Reference
15	1-Phenyl-3 (p-methoxylphenyl) triazene-1-oxide		Anti-inflammatory	Miesel JL et al. (1976)
16	1-Phenyl-3 (mo-florophenyl) triazene-1-oxide		Anti-inflammatory	Miesel JL et al. (1976)

medicinally useful compounds. In general, it seems a less studied area of medicinal chemistry. The author's group has, however, attempted some of the metal complexes for their medicinal/bioactivity, but very few complexes have been explored to date. There is a serious need to synthesize some metal complexes of Pt and Pd in particular and screen their biological or medicinal applications. The author's research group has screened more than 200 hydroxytriazenes for their different activities, but very few metal complexes for this use. To highlight this and to suggest a future strategy for researchers in this excellent field, a table of compounds with their available characterization data and activity is reported. The work done on the class named as triazene-1-oxides is also summarized to attract attention to the volume of work done on this class of compounds (Tables 3.3 and 3.4).

References

1. Stevens, M. F. G., and Newlands, E. S. From triazines and triazenes to temozolomide. *Eur. J. Cancer* 1993, 29A(7), 1045.
2. Agarwal, S. S., and Kirkwood, J. M. Temozolomide, a novel alkylating agent with activity in the central nervous system, may improve the treatment of advanced metastatic melanoma. *Rev. Oncol.* 2000, 5, 144.
3. Quirbt, I., Verma, S., Petrella, T., Bak, K., and Charette, M. Temozolomide for the treatment of metastatic melanoma. *Curr. Oncol.* 2007, 14, 27.
4. Marchesi, F., Turriziani, M., Tortorelli, G., Avvisati, F., Torino, F., and Vecchis, L. D. Triazene compounds: Mechanism of action and related dna repair systems. *Rev. Pharmacol. Res.* 2007, 56, 275.
5. Rachid, Z., Katsoulas, A., Brahimi, F. Jean, Claude. Synthesis of pyrimidinopyridine-triazene conjugates targeted to abl tyrosine kinase. B. J. *Bioorg. Med. Chem. Lett.* 2003, 13, 3297.
6. Yahalom, J., Voss, R., Leizerowitz, Z., and Polliack, A. Secondary leukemia following treatment of Hodgkin's disease: Ultrastructural and cytogenetic data in two cases with a review of the literature. *Am. J. Clin. Pathol.* 1983, 80, 231.
7. O'Reilly, S. M., Newlands, E. S., Glaser, M. G., Brampton, M., Rice Edwards, J. M., and Illingworth, R. D. Cytotoxic effect of interferon on primary malignant tumour cells. Studies in various malignancies. *Eur. J. Cancer* 1993, 29, 1940.
8. Unsalam, S., Cikla, P., Kucukguzel, S. R., Sahin, F., and Bayark, O. Synthesis and characterization of triazenes derived from sulfonamides. *Marmara. Pharm. J.* 2011, 15, 11
9. Friedman, H. S., Kerby, T., and Calvert, H. Temozolomide and treatment of malignant glioma. *Rev. Clin. Cancer Res.* 2000, 6, 2585.
10. Fukushima, T., and Takeshima, H. Anti-glioma therapy with temozolomide and status of the DNA-repair gene MGMT. *Anticancer Res.* 2009, 29, 4845.
11. Braithwacle, A. W., and Baguley, B. C. Existence of an extended series of antitumor compounds which bind to deoxyribonucleic acid by nonintercalative means. *Bio-Chem.* 1980, 19, 1101.
12. Pilch, D. S., Kirolos, M. A., Liv, X., Plum, G. E., and Breslauer, K. Berenil binding to higher ordered nucleic acid structures: Complexation with a DNA and RNA triple helix. *J. Bio-Chem.* 1995, 34, 9962.

13. Sava, G., Giraldi, I., Lassiani, L., and Nisi, C. Antimetastatic action and hematological toxicity of p-(3,3-dimethyl-1-triazeno)benzoic acid potassium salt and 5-(3,3-dimethyl-1-triazeno)imidazole-4-carboxamide used as prophylactic adjuvants to surgical tumor removal in mice bearing B16 melanoma. *Cancer Res.* 1984, 44, 64.
14. Sava, G., Giraldi, I., and Lassiani, L. Effects of isomeric aryldimethyltriazenes on Lewis lung carcinoma growth and metastases in mice. *Chem. Biol. Interact.* 1983, 46, 131.
15. Giraldi, T., Sava, G., Cuman, R., Nisi, C., and Lassiani, L. Selectivity of the antimetastatic and cytotoxic effects of 1-p-(3,3-dimethyl-1-triazeno)benzoic acid potassium salt (+/-)-1,2-di(3,5-dioxopiperazin-1-yl)propane, and cyclophosphamide in mice bearing Lewis lung carcinoma. *Cancer Res.* 1981, 41, 2524.
16. Kažemėkaitė, M., V. Šimkevičienė, and Z. Talaikytė. Antileucemic activity of 4-(3-dimethyltriazeno)-N-acylbenzenesulfonamides. *Biologija, Priedas,* 2000, 2, 242.16
17. Larsen, J. S., Zahran, M. A., Pedersen, E. B., and Neelsen, C. Synthesis of triazenopyrazole derivatives as potential inhibitors of HIV-1. *Montash. Chem.* 1999, 130, 1167.
18. O'Reilly, S. M., Newlands, E. S., Glaser, M. G., Brampton, M., Rice Edwards, J. M., Tllingworth, R. D., Richards, P. G., Kennard, C., Colquohoun, I. R., Lewis, P., and Stevens, M. F. G. Temozolomide: A new oral cytotoxic chemotherapeutic agent with promising activity against primary brain tumours. *Eur. J. Cancer* 1993, 29A, 940.
19. Blechen, N. M., Newlands, E. S., Lee, S. M., Thatcher, N., Selby, P. Calvert, A. H., Rustin, G. j., Brampton, M., and Steven, M. F. Cancer research campaign Phase II trial of temozolomide in metastatic melanoma. *J. Clin. Oncol.* 1995, 13, 910.
20. Hoang-Xuan, K., Camillen-Broet, S., and Soussan, C. Recent advances in primary CNS Lymphoma. *Curr. Opin. Oncol.* 2004, 16, 601.
21. Tani, M., Fina, M., Alinari, L., Stefoni, V., Baccarani, M., and Zinzani, D. L. Phase II trial of temozolomide in patients with pretreated cutaneous T-cell lymphoma. *Haeinato Logica* 2005, 90, 1283.
22. Kanjeekal, S., Chambera, A., Fung, M. F., and Verma, S. Systemic therapy for advanced uterine sarcoma: A systematic review of the literature. *Gynecol. Oncol.* 2005, 97, 624.
23. Tollenaere, J. P. The role of structure-based ligand design and molecular modelling in drug discovery. *Pharm World Sci.* 1996, 18(2), 56.
24. Yu, Hongshi, and Adedoyin, A. ADME-Tox in drug discovery: Integration of experimental and computational technologies. *Drug Discov. Today* 2003, 8(18), 852.
25. Agarwal, Shilpa, Barolya Bhargava, Amit, Tripathi, I. P., and Goswami, A. K. Synthesis, characterization, theoretical prediction of activities and evaluation of biological activities of some sulfacetamide based hydroxytriazenes. *Bioorg. Med. Chem. Lett.* 2016, 26, 2870.
26. Stepanchikova, A. V., Lagunin, A. A., Filmonov, D. A., and Poroikov, V. V. Prediction of biological activity spectra for substances: Evaluation on the diverse sets of drug-like structures. *Curr. Med. Chem.* 2003, 10, 225.
27. Anzalil, Barniekel, B. Cezzane, Krug, M., Filmonov, D., and Poroikvo, V. Discriminating between drugs and nondrugs by prediction of activity spectra for substances (PASS). *J. Med. Chem.* 2001, 44, 2432.

28. Van de Waterbeemd, H., and Gifford, Eric. ADMET *in-silico* modelling: Towards prediction paradise? *Nature* 2003, 192–204. www.nature.com/reviws/drugdisc.
29. Borisov, E. V., Zykova, T. N., and Yashunskii, V. Some 3-hydroxytriaznes derivatives and their antitubercular activity. *Zhu. Obsch. Khim.* 1966, 36, 2125.
30. Gubler, K. Insecticidal and acaricidal 1-phenyl-3-hydroxytriazenes. *Ger. Offen* (CA 1970, 73, 13181C) 1970; 2003333-54.
31. John, L. M. U.S. Patent, 3,3-Dialkyl-1-(substituted-phenyl) triazene-1-oxides. USUS3989680A.
32. Ajay, K. Goswami and Purohit, Dau N. Synthesis and antimicrobial activities of some hydroxytriazenes: A new class of biologically active compunds. *Anal. Sci.* 2001, 17, 1789.
33. Goswami, A. K. Syntehsis and tusecticidal activity of some hydroxytriaznes and their vandium complexes. *Pest. Res. J.* 2002, 14(2), 213.
34. Hura, I. S., Naulakha, Neelam, Goswami, A. K., and Srivastava, M. K. Biological activity of some substituted hydroxytriazenes. *Indian J. Microbiol.* 2003, 43(4), 275.
35. Chauhan, L. S., Jain, C. P., Chauhan, R. S., and Goswami, A. K. Wound healing activity of some substituted hydroxytriazenes. *Adv. Pharmacol. Toxicol.* 2006, 7(3), 73.
36. Chauhan, Lalit Singh, Jain, C. P., Singh, Chetan, Chauhan, R. S., and Goswami, A. K. Anti-inflammatory activity of some substituted hydroxytriazenes on wister raly. *Bio Sci. Biotechnol. Res. Asian* 2006, 3(2), 381.
37. Chauhan, L. S., Jain, C. P., Chauhan, R. S., and Goswami, A. K. Wound healing activity of some substituted hydroxytriazenes. *Adv. Pharmacol. Toxicol.* 2006, 7(3), 73.
38. Chauhan, L. S., Jain, C. P., Chauhan, R. S., and Goswami, A. K. Synthesis and halgesic activity of hydroxytriazenes. *Asian J. Chem.* 2007, 19(6), 4684.
39. Singh, Kalpana, Patel, Peeyush, and Goswami, A. K. Anti-inflammatory activity of hydroxytriazenes and their vanadium complexes. *J. Chem.* 2008, 5(52), 1144.
40. S. Kumar, Garg, Meenakshi, Jodha, J. S., Singh, R. P., Pareek, Nedam, Chauhan, R. S., and Goswami, A. K. Studies on insecticide activity of some hydroxytriazenes derivatives. *J. Chem.* 2009, 6(2), 466.
41. Neelam, Nawakha, Meenakshi, Garg, Jodha, J. S., Chundawat, N. S., Patel, Peeyush, Chauhan, R. S., and Goswami, A. K. Synthesis and insecticidal activity of some substituted hydroxytriazenes. *Pestology* 2008, 32(9), 48.
42. Neelam, Nawlakha, Garg, Meenakshi, Chauhan, R. S., and Goswami, A. K. Antifungal activity of hydroxytriazenes and their Cu(II) complexes. *Pestology* 2009, 33(2), 46.
43. Chauhan, L. S., Jain, C. P., Chauhan, R. S., and Goswami, A. K. Wound healing activity of hydroxytriazenes. A new class of bio-active e-compounds. *J. Chem. Param. Res.* 2010, 2(1), 539.
44. Chauhan, L. S., Jain C. P., Chauhan, R. S., and Goswami, A. K. Synthesis of some substituted hydroxytriazenes and their analgesic activity. *J Chem. Parm. Res.* 2010, 2(4), 999.
45. Joshi, P., Pareek, N., Upadhyay, D., Khanam, R., Bhandari, A., Chauhan, R. S., and Goswami, A. K. Polarographic determination and antifungal activity of Cu(II) complex with 3-hydroxy-3-m-tolyl-1-p-(Sulphonamide) phenyltriazene. *Int. J. Pharm. Sci. Drug Res.* 2010, 2(4), 278.

46. Chauhan, L. S., Jain, C. P., Chauhan, R. S., and Goswami, A. K. Synthesis and antimicrobial activity of some substituted hydroxytriazenes. *J. Chem. Pharm. Res.* 2010, 2(4), 979.
47. Gokargu, Ganga Raju, Kasina, Sudhakar, Gokargagu, Raju, Rama, Golakoti, Trimurtula, Some paili, Venkateshwaralu; Krishanu, Sengupta, and Bhupathiraju, Kiran. Preparation of triazene analogs useful in treatment of metastatic malignant melanoma and oliieer cancers. *PCT Int. Appls.* 2010 W02010029 577A2 20100318.
48. Domingues, Vanessa O., Horner, Rosmari, Reetz, Luz G. B., Kuhn, Fabio, Coser, Virgina M, Rodrigues Jacqueline, N, Bauchspiess, Rita, Periera, Walder V., Paraginski, Gustavo L., Locatelli, Aline, et al. In vitro evaluation of triazenes: DNA cleavage antibacterial activity and cytotoxicity against acute myeloid leukemca cells. *J. Brazil. Chem. Soc.* 2010, 21(2), 2226.
49. Osmak, Maja, Polanc, Slovenko, Cimbora, Zovko, Tamara, Brozovik, Anamaria, Kocevar, Marijan, Majce, and Vita, Allics Branco. *PCT Int. Appl.* 2010, WO 20101033338 A1 201000916.
50. Chauhan, L. S., Jain, C. P., Chauhan, R. S., and Goswami, A. K. Toxicity studies of substituted hydroxytriazenes on rodents. *J. Glob. Pharma. Technol.* 2010, 2(11), 11.
51. Dlana, Patrizia, Stagno, Antonina, Paola, Barraja, Carbone, Anna, Parrino, Barbara, Dall Acqua, Francesco, Vedaldi, Deniela, Salvador, Alessia, Brun, Paola, Castaguuolo, Igarizio, et al. Synthesis of triazenoaza indoles: A new class with antitumor activity. *Chem. Med. Chem.* 2011, 6(7), 1291.
52. Ombaka, O., and Cichumbi, J. M. Synthesis and insecticidal activities of some selected hydroxytriazenes. *J. Env. Chem. Ecotoxicol.* 2011, 3(11), 286.
53. Unsalan, Seda, Cikla, Pelin, Kucukguzel, S. Guniz, Rollas, Sevin, Sahin, Fikrettin, Bayrack, Faruk Omar. Synthesis and characterization of triazenes derived from sulfonamides. *Marm. Pharm. J.* 2011, 15(11), 11.
54. Cimbora-Zovko, Tamara, Brozovic, Anamaria, Plantanida, Ivo, Fritz, Gerhard, Virag, Andrej, Alic, Branko, Majce, Vita, Kocevar, Marijan, Polanc, Slovenko, and Osmak, Maja. Synthesis and biological evaluation of 4-nitro-substituted 1,3-diaryltriazenes as a novel class of potent antitumor agent. *Eur. J. Med. Chem.* 2011, 46(7), 297.
55. Chauhan, L. S., Jain, C. P., Chauhan, R. S., and Goswmai, A. K. Synthesis and neuropharma idogoical effects of some substituted hydroxytriazenes. *J. Glob. Pharma. Technol.* 2011, 3(6), 9.
56. Rehana, K., Kishore, S., Saba, K., and Goswami, A. K. Synthesis characterization and antifungal activity of 3-hydroxy-3-p-tolyl-1-m-nitrophenyltriazene and its complex with iron (III). *J. Chem. Pharm. Res.* 2011, 3(6), 57.
57. Babel, Tushita, Bhandari, Amit, Jain, Preksha, Mehta, Anita, and Goswami, A. K. Synthesis activity prediction and spectrophotometric study of molybednum complex of 3-hydroxy-3-p-tolyl-1-p-carboxyphenyltriazene. *Int. Res. J. Pharma* 2012, 3(5), 382.
58. Bhandari, Amit, Rudra, Singh P., Sharad, Sharma, Mehta, Anita, and Goswami, A. K. Antimicrobial evaluation of some hydroxytriazenes and their ternary metal complexes. *IJPRT* 2012, 4(11), 55.
59. Singh, R. P., Bhandari, Amit, Bhatt, Ritu, Sharma, Sharad, Mehta, Anita, Chauhan, R. S., and Goswami, A. K. Biological activity evluation of some activity hydroxytriazenes. 2012, *Int. J. Chem.* 1(3), 332.

60. Baroliya, Prabhat K., Regar, Mangilal, Chauhan, R. S., and Goswami, A. K. Synthesis characterization and antimicrobial activities of hydroxytriazenes and their Co(II) complexes. *AFFINIDAD*, 2014, LXXXI, 568, 305.
61. Khanum, Rehana, Jodha, J. S., Dashora, R., and Goswami, A. K. Antifungal activity of hydroxytriazenes and their ternary complexes with vanadium (V) and thiourea. *J. Glob. Pharma Technol.* 2014, 6(11), 1.
62. Rehana, Khanam, Jodha, J. S., Dashora, R., and Goswami, A. K. Synthesis and insecticidal activity of some hydroxytriazenes derivatives and their ternary complexes with vanadium (V) and thiourea. *J. Glob. Pharma Technol.* 2014, 6(12), 1.
63. Chundawat, Narendra Singh, Pandya, Mayank, Pal Singh, Girdhar, and Chauhan, R. S. Antibacterial activity of hydroxytriazenes, Schiff's base and their ternary complexes of zinc (II) with Schiff's base and hydroxytriazenes. *World J. Pharm. Pharm. Sci.* 2015, 4(2), 330.
64. Agarwal, S., Baroliya, P. K., Bhargava, A., Tripathi, I. P., and Goswami, A. K., Synthesis characterization theoretical prediction of activities and evaluation of biological activities of some sulfacetamide based hydroxytriazenes. *Bioorg. Med. Chem. Lett.* 2016, 26, 2870.
65. Regar, Mangilal, Baroliya, P. K., Patidar, Ashok, Dashora, Rekha, Mehta, Anita, Chauhan, R. S., and Goswami, A. K. Antidyslipedemic and antioxidant effects of novel hydroxytriazenes. *Pharm. Chem. J.* 2016, 50(5), 310.
66. Goswami, A. K. Sharma, P., Agaral, S., and Khan, I. Hydroxytriazenes: A promising class of antiinflammatory compounds. *MOJ. Org. Chem.* 2017, 1(3), 1.

chapter four

Triazenes and triazines

Nitrogen-containing compounds, both cyclic (triazines) and acyclic (triazenes), have crucial roles in medicinal chemistry and pharmaceutical research. In this chapter we will discuss the chemistry of triazenes and triazines and also highlight some of the recent advances and diversity possible with these types of structural frameworks. The acyclic triazenes (RN=N–NR'R''), also known as triazanylene, are a class of compounds having much promise in synthetic chemistry as they are reactive groups which are both stable and adaptable to numerous synthetic transformations including total synthesis, polymer technology, and the construction of novel heterocycles. Triazenes also have a tendency to surprise medicinal chemists with new reaction pathways and increasing applicability. Triazenes have been studied for their anticancer potential [1, 2], and used as protecting groups in natural product synthesis [3], combinatorial chemistry [4], polymer synthesis [5], oligomer synthesis [6], and heterocycle synthesis [7].

The cyclic triazine is a heterocyclic system, analogous to the six-membered benzene ring with alternate three carbons and three nitrogens. The three isomers of triazine are distinguished from each other by the positions of their nitrogen atoms, and are referred to as 1,2,3-triazine, 1,2,4-triazine, and 1,3,5-triazine. The cyclic triazine derivatives have been widely used in several fields as herbicidals [8,9], antimicrobials, antituberculars, anticancers, antivirals, and antimalarials [10]. Melamine derivatives have also found applications like host–guest [11] or super-structure assemblies [12]. The chapter is divided in two parts, as we explore the synthesis and medicinal applications of triazenes and triazines.

4.1 Triazenes

In the implementation of combinatorial chemistry in the modern drug discovery process, the approach to novel diverse heterocycle libraries is an indispensable requirement. Triazenes, which are concealed diazonium salts, can be used to link functionalized arenes and amines to generate various heterocyclic structures. Since triazene anchors are stable toward various reagents and perform well under a range of reaction conditions, these multifunctional linkers are well-suited for the syntheses of complex organic molecules, such as natural products.

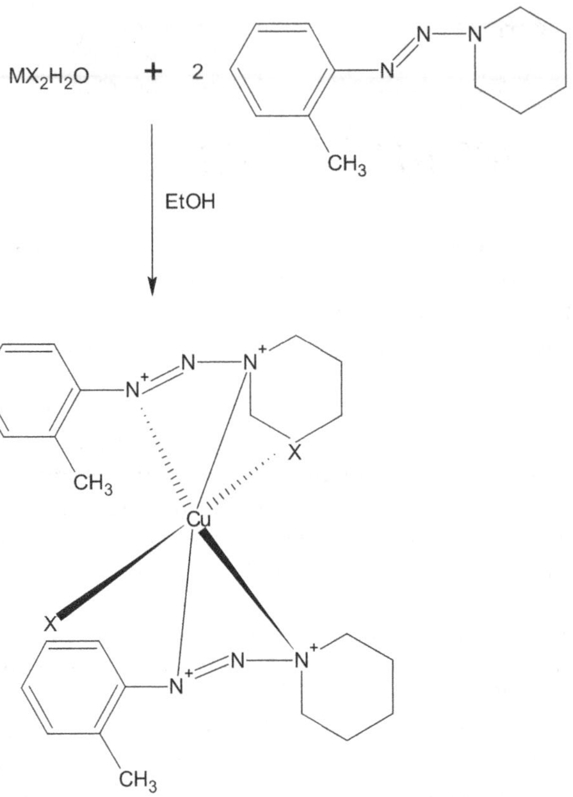

Scheme 4.1 Preparative methods for complexes (1:2) molar ratio (M:L).

A new series of complexes containing triazene compounds was developed by Nuha H. Al-Saadawy et al. [13] by the diazotization of *o*-toluidine followed by the treatment of piperidine to give triazene compound (1-(*o*-tolyldiazenylpiperidine) (Scheme 4.1). The complexes $M[(2-CH_3-C_6H_3-N=N)-N-C_5H_{10})]_2X_2-$ of Cu(II), Ni(II), Co(II), Zn(II), and Fe(II) with triazene compound (1-(o-tolyldiazenyl)piperidine) have been synthesized by the reaction of copper bromide, hydrate nickel chloride, hydrate cobalt chloride, zinc chloride, and iron bromide in a 1:2 mole ratio. Several new coordination compounds containing triazene were prepared in this study. The FTIR, CHN, and 1H NMR studies of the complexes showed significant evidence in favor of complexation 1:2 (M:R), when the complexes were prepared via refluxation.

Carolina Torres-García et al. [14] developed a novel method for the synthesis of para-substituted phenylalanine-containing cyclic peptides. This strategy involves the coupling of phenylalanine to the solid support through its side chain *via* triazene linkage, on-resin cyclization of the

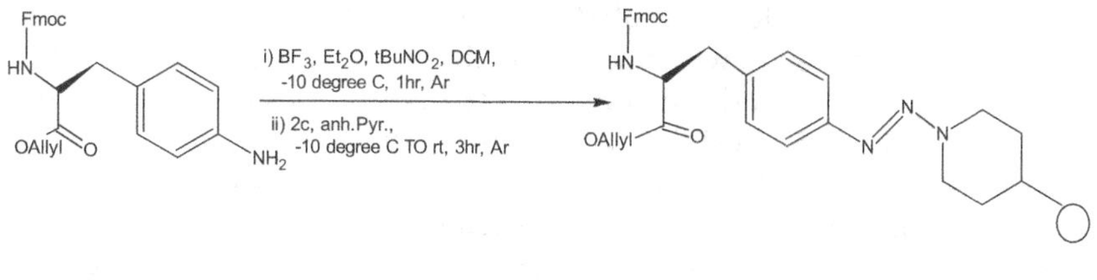

Scheme 4.2 Coupling of protected phenylalanine to the piperidine resin *via* triazene linkage.

peptide chain, cleavage of the cyclic peptide from the resin under mild acidic conditions, and further transformation of the resulting diazonium salt (Scheme 4.2). The important feature of this approach is exemplified by the solid-phase synthesis of some derivatives of the naturally occurring cyclic depsipeptidezygosporamide.

In this work, Carolina Torres-García et al. developed a new versatile and efficient protocol for the solid-phase synthesis of phenylalanine containing peptides, based on the anchorage of the aromatic ring of the side chain to the solid support through a triazene linkage. The use of a low percentage of TFA for peptide cleavage from the resin and further chemical transformation of the resulting diazonium salt under suitable experimental conditions allows the introduction of diversity at the para position of the aromatic ring. It is remarkable that this strategy can also be used for the generation of cyclodepsipeptide derivatives due to the mild conditions used, which preserve the integrity of the ester bond. Furthermore, the high versatility in terms of commercial availability of building blocks for peptide synthesis (orthogonal protecting groups) makes this strategy compatible with the presence of trifunctional amino acids in the peptidic sequence. In this case, the side chain protecting groups should be removed after pH-derivatization.

Jonas Lohse et al. [15] developed a method for identifying triazene linker that helps identify azidation sites of labeled proteins *via* a click and cleave strategy. Azide-bearing molecules are immobilized on functionalized sepharose beads *via* copper-catalyzed Huisgen-type click chemistry and selectively released under acidic conditions by chemical cleavage of the triazene linkage. They applied this method to identify the modification site of targeted diazotransfer on BirA (Scheme 4.3).

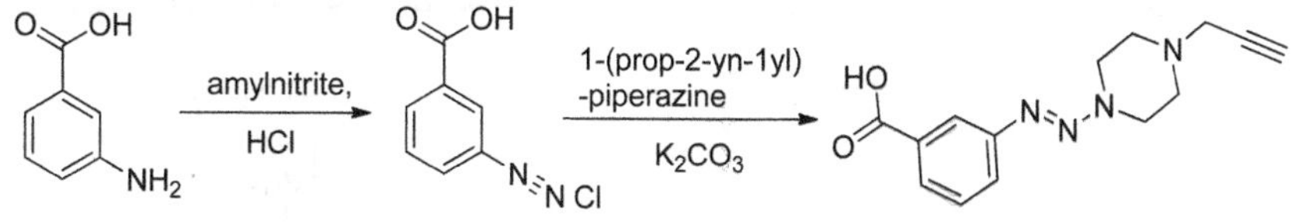

Scheme 4.3 Synthesis of the clinker resin.

Jonas et al. also demonstrated that the immobilization of azide-bearing species with the newly developed clinker resin is possible for small molecules, peptides, and proteins. Furthermore, they established two new "target and identify" workflows that allowed the identification of the regioselective diazotransfer modification sites. The modified amino acid of DtBio-labeled streptavidin and the novel target BirA could be respectively identified after CuAAC on the protein level and on-bead digestion and after performing the click reaction on the peptide level, that is, the azidopeptide could be enriched from a BirA in-gel digest.

A continuous-flow protocol for the synthesis of triazenes, followed by diazonium salts generation and subsequent conversion into their masked or protected triazene derivatives using secondary amine has been reported by Christiane Schotten et al. [16] (Scheme 4.4). The process has also been applied to prepare the antineoplastic agents mitozolomide and dacarbazine.

Seda Unsala et al. [17] synthesized a series of novel triazene derivatives *via* coupling of diazonium salts of amines (sulfaguanidine, sulfapyridine, sulfamethoxazole) with N-methylaniline/p-nitroaniline in acidic medium. The structures of the synthesized compounds were confirmed by the spectroscopic data (UV, IR, 1H-NMR, APCI-MS) and elemental analysis (Scheme 4.5).

They tested the synthesized compounds for their anticancer activities and cytotoxicity properties using doxorubicin and taxol as standard drugs. A 549 and L 929 cell lines were used to test both anticancer effects and cytotoxicity. The cytostatic activity discovered for several substituted triazenes has stimulated intense research efforts directed toward the development of triazene-based drugs.

An interesting application of triazenes, explored by Jurgen Stebani et al. [18], is based on the fact that the triazeno group is comparatively stable with respect to thermal decomposition, but undergoes facile photochemical cleavage, releasing nitrogen. This property may lead to novel applications in photolithography and photo reproduction.

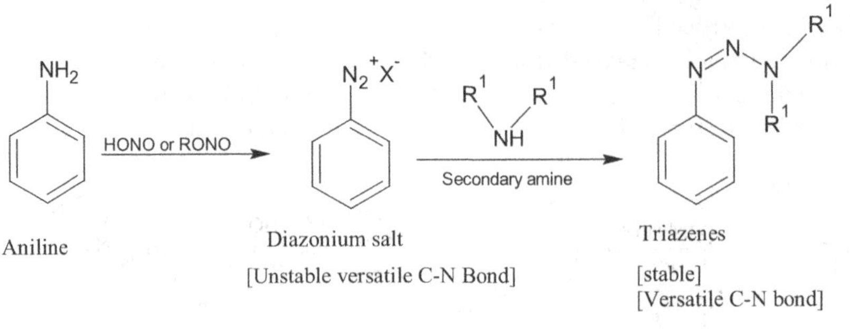

Scheme 4.4 Preparation of triazenes from diazonium compounds.

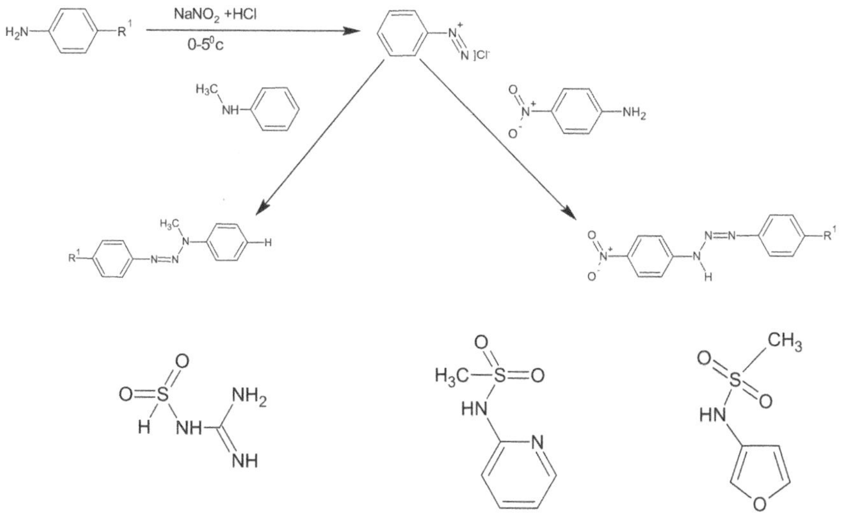

Scheme 4.5 The synthesis pathway of the studied compounds (sulfaguanidine, sulfapyridine, sulfamethoxazole substituted).

Jurgen Stebani et al. also investigated the synthesis of a novel photopolymer which contains two triazeno groups per repeating unit (Scheme 4.6). This synthesis was carried out *via* a polycondensation reaction in which a bis-diazonium salt and a bifunctional secondary aliphatic amine react in an interfacial polycondensation yielding a polymer which contains the photosensitive triazene group in the polymer backbone.

Margaret L. Gross et al. [19] have used the triazene moiety as a protecting group for aromatic amines during the generation of an aromatic carbanion elsewhere in the ring, that is, by metal halogen exchange and during the subsequent reaction of this carbanion with a variety of electrophiles, and they explored a reaction of (2-, (3-, and (4-bromophenyl)-3,3-(1,4-butanediyl) triazene (derived from the corresponding 2-, 3-, and 4-bromoanilines) with sec- or tert-butyllithiums afforded the corresponding (triazenylaryl1)-lithiums; these intermediates reacted with a variety of electrophiles to give substituted aryl triazenes. These products could be converted to substituted anilines by reduction or to novel substituted aromatics by a modified Sandmeyer reaction. Also, Gross et al. anticipated that no directing effect would be likely due to the diazo nature of the derivatized aniline nitrogen atom, products in this case being derived from unstabilized mono anions. The potential utility of this group was further enhanced by the fact that triazenes have been converted to their corresponding anilines by reduction under mild conditions.

In a study carried out by Weronika Pawelec et al. [20], triazene compounds have emerged as a novel and effective class of flame retardants

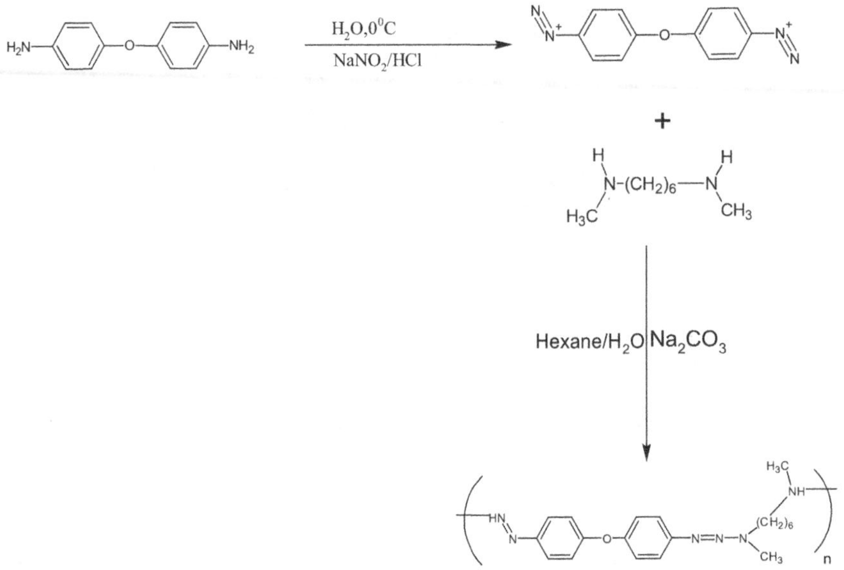

Scheme 4.6 Synthesis of triazene polymer.

for polypropylene. In this study, they synthesized four triazene derivatives, that is, bis-4,4′-(3,3′-dimethyltriazene)-diphenyl ether (1), bis-4,4′-(3,3′-diethyltriazene)-diphenyl ether (2), 2,2,6,6,-tetramethyl-1-phenylazopiperidine (3), and 4-hydroxy-2,2,6,6-tetramethyl-1-phenylazopiperidine (4) and determined their thermal properties by differential scanning calorimetry (DSC), and the fragmentation patterns were analyzed by simultaneous mass spectrometry (MS) and Fourier transform infrared (FTIR) spectrometry of off-gases from a thermogravimetric analyzer (TGA) (Scheme 4.7).

These triazenes exhibited an exothermic decomposition peak at temperatures between 230 and 280°C when the triazene units were homolytically cleaved into various aminyl, resonance-stabilized aryl radicals, and different CH fragments with the simultaneous evolution of elemental nitrogen.

The potential of triazenes as a new class of flame retardants for polypropylene films was studied by performing an ignitability test in accordance with DIN 4102-1/B2 standard. Polypropylene samples containing very low concentrations of only 0.5 wt% of any of these triazene (ReN1 ¼ N2eN3R′R00) additives passed the test with B2 classification. Notably, no burning dripping could be detected. The average burning times are very short with exceptionally low weight losses. Based on this preliminary FR testing they have shown that the triazene compounds constitute a new and interesting family of radical generators for the flame retarding of polymeric materials.

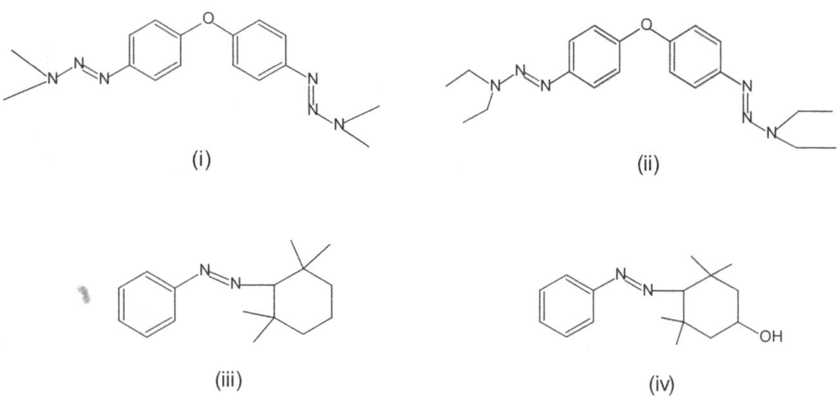

Scheme 4.7 Molecular structures of triazenes.

A series of 3-aminoacyl-1-aryl-3-methyltriazenes was synthesized by Emilia Carlvalho et al. [21] by the reaction of 1-aryl-3-methyltriazenes with N-BOC protected amino acids using the DCC method of activation, followed by deprotection of the amino functional group using HCl in nitromethane (Scheme 4.8). Half-lives for the hydrolysis of these compounds to the corresponding monomethyl-triazenes at 37°C in isotonic phosphate buffer and in 80% human plasma containing 20% phosphate buffer were determined by HPLC.

In another study, Emilia Carvalho et al. [22] synthesized a series of 3-acyloxymethyloxycarbonyl-1-aryl-3-methyltriazenes by the sequential reaction of 1-aryl-3-methyltriazenes with chloromethyl chloroformate and NaI in dry acetone (Scheme 4.9). They have also explored the synthesis strategy using silver carboxylate. The hydrolysis of these compounds was studied in pH 7.7 isotonic phosphate buffer and in human plasma. Triazene acyloxycarbamates demonstrated their ability to act as substrates for plasma enzymes. Emilia observed that the reaction is also buffer catalyzed. Thus, at pH 7.7, pH-independent, base-catalyzed, and buffer-catalyzed processes all contribute to the hydrolysis reaction. The sensitivity of the hydrolysis reaction to various structural parameters in the substrates indicates that hydrolysis occurs at the ester rather than the carbamate functionality. This study indicates that aminoacyltriazenes are prodrugs of the antitumor monomethyl-triazenes. They combine chemical stability with rapid enzymatic hydrolysis, and are consequently good candidates for further prodrug development. Moreover, this type of derivative allowed the synthesis of mutual prodrugs, associating the antitumor monomethyl-triazenes with anti-inflammatory NSAIDs as well as with the anticancer agent butyric acid.

A series of 1,4-di-[2-aryl-1-diazenyl]-*trans*-2,5-dimethylpiperazines has been synthesized by Vanessa Renee Little et al. [23] using

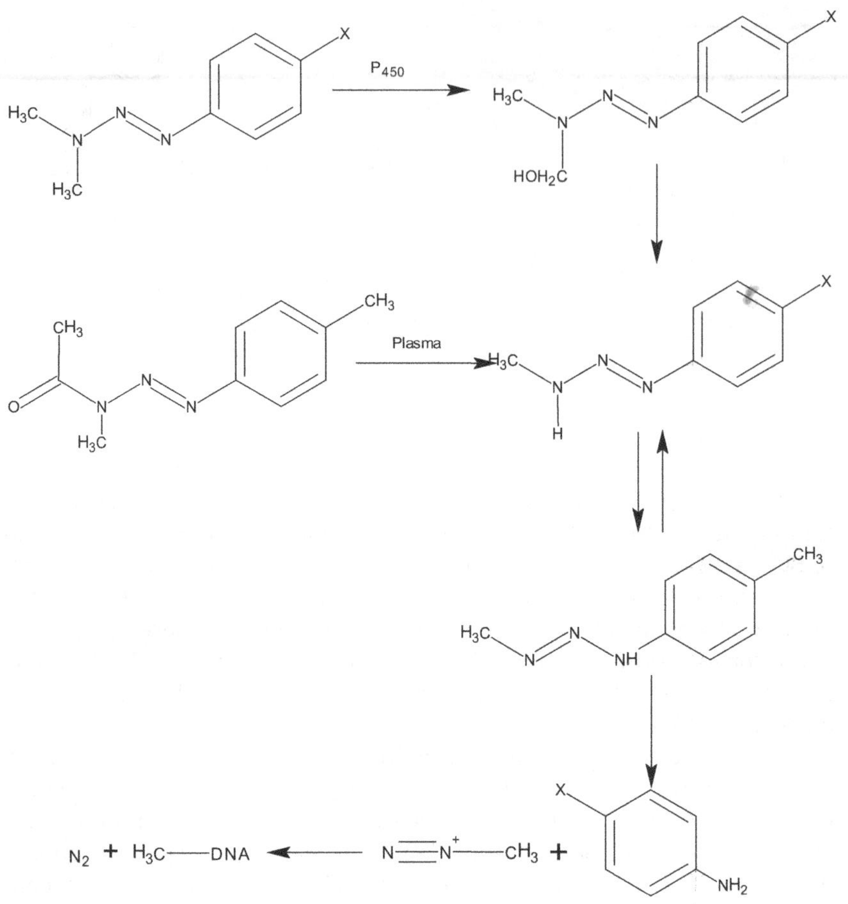

Scheme 4.8 Pathways for the metabolism of both 1-aryl-3,3-dimemethyltriazenes and aminoacyltriazenes.

trans-2,5-dimethylpiperazine with two equivalents of the appropriate diazonium salt. The presence of stereocenters at C2 and C5 of the pipera-zine ring in the bis-triazene creates two unique pairs of diastereotopic protons in the methylene groups at positions 3 and 6 of the piperazine ring (Figure 4.1).

The depolymerization of a triazeno group containing photopolymer, poly[oxy-1,4-phenyleneazo(methylimino)hexamethylene(methylimino) azo-1,4-phenylene], was studied in a tetrahydrofuran (THF) solution by Oskar Nuyken et al. [24]. Irradiation of the material leads to a clean decomposition of the photolabile polymer, as monitored by UV spec-troscopy and gel-permeation chromatography (GPC). As compared to the photolysis in THF solution, the light-induced decomposition rate of a

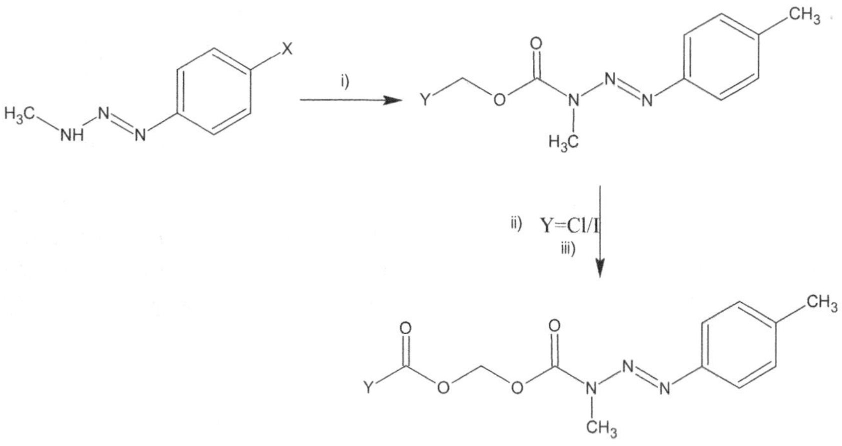

Scheme 4.9 Synthesis of triazenes. Reagents: (i) ClCH₂OCOCl; (ii) NaI; (iii) RCO₂Ag or RCO₂H/Ag₂CO₃.

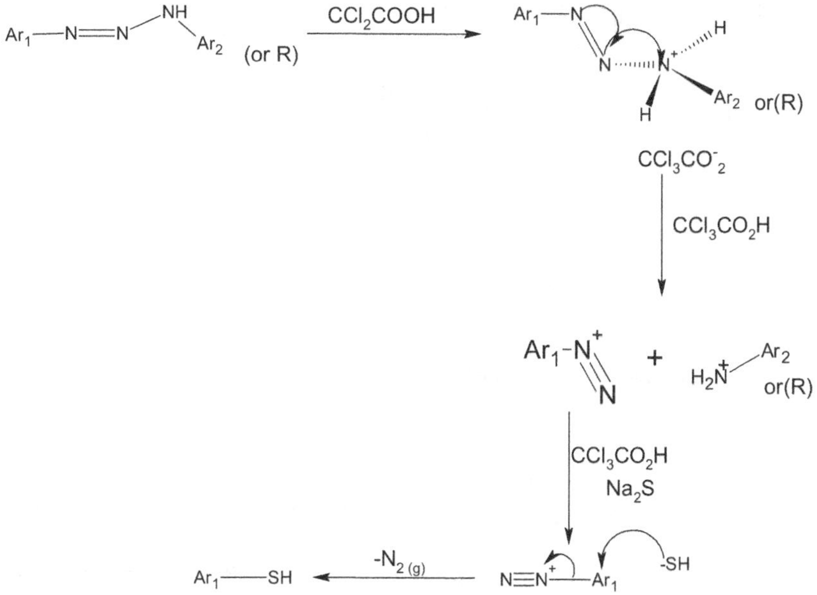

X= p-CN / p-NO₂/p-CO₂Me/ p-CO₂Et/ p-COCH₃/ p- Cl/p-Br/ -CH₃ /p-OCH₃ / H/ O-CN/ O-NO₂ /O-Br

Figure 4.1 Structure of 1,4-di-[2-aryl-1-diazenyl]-*trans*-2,5-dimethylpiperazines.

polymer in a polymer film is shown to be much slower. The highly photosensitive triazeno group also decomposes thermally at temperatures
above approx. 220°C (Figure 4.2). The kinetics of thermal degradation
of the polymer in substance was investigated at a temperature of 256°C,
and monitored by GPC measurements. During this decomposition, they
observed the development of higher molar-mass fractions, which result
from grafting reactions of primary radicals. Upon further thermolysis the
triazene polymer is completely degraded to low-molar-mass products.
The volatile decomposition products were identified by gas chromatogra-
phy/mass spectrometry (GC/MS) analysis. The protolytic decomposition,

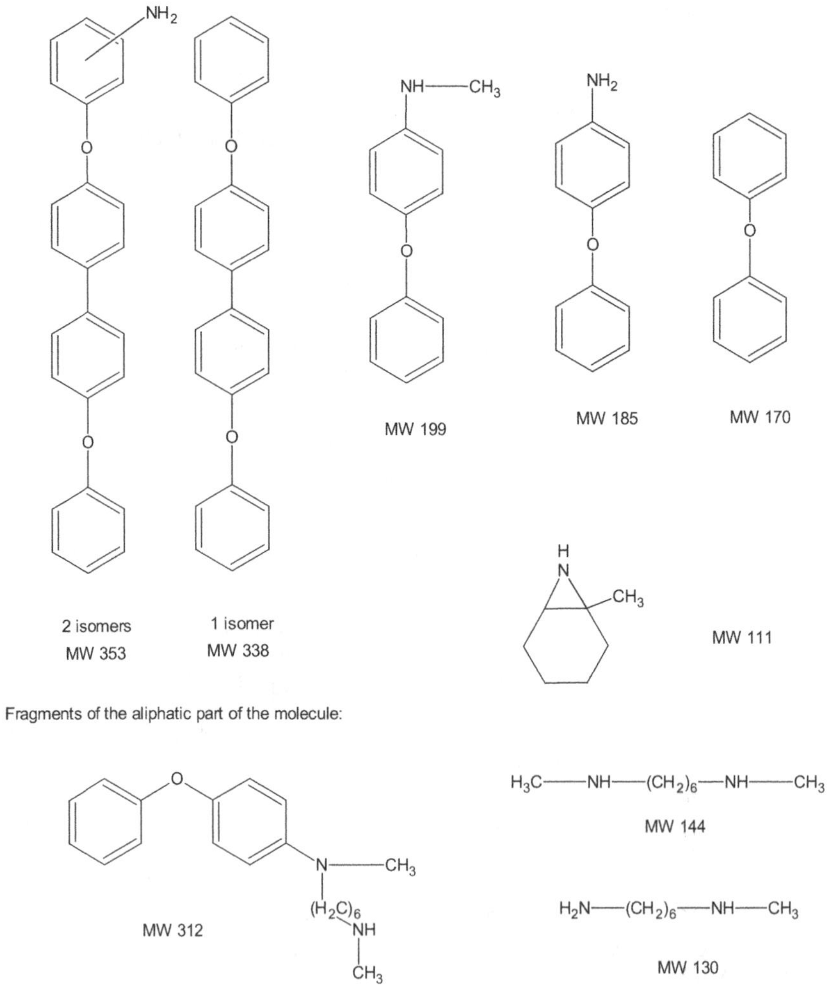

Figure 4.2 Products of the thermolysis identified by GC/MS.

which represents the retrosynthesis of the triazene polymer, was studied in a 9:1 mixture of tetrahydrofuran and an aqueous citrate buffer solution. Although the decomposition rate in this solvent mixture is slow, as compared to the depolymerization in diluted hydrochloric acid, a clean decomposition of the triazene polymer is obtained.

Chengming Wang et al. [25] developed a general synthesis of unprotected indoles through a triazene-directed C–H annulation using alkynes. This reaction is proposed to undergo a 1,2-rhodium shift ring contraction or an N=N insertion mechanism. Excellent regioselectivity was achieved by asymmetrically substituted alkynes, which found immediate synthetic applications (Scheme 4.10).

Triazenes can be used beneficially to link functionalized arenes and amines which was studied by Stefan Brase [26] to generate various heterocyclic structures, particularly benzoannelated nitrogen heterocycles imbedding classical and unusual yet useful nitrogen rings upon cleavage from the resin (Scheme 4.11). They therefore offer flexibility in the synthesis of the diverse heterocycle libraries needed for a successful drug discovery program. Since triazene anchors are stable toward various reagents and perform well under a range of reaction conditions, these multifunctional

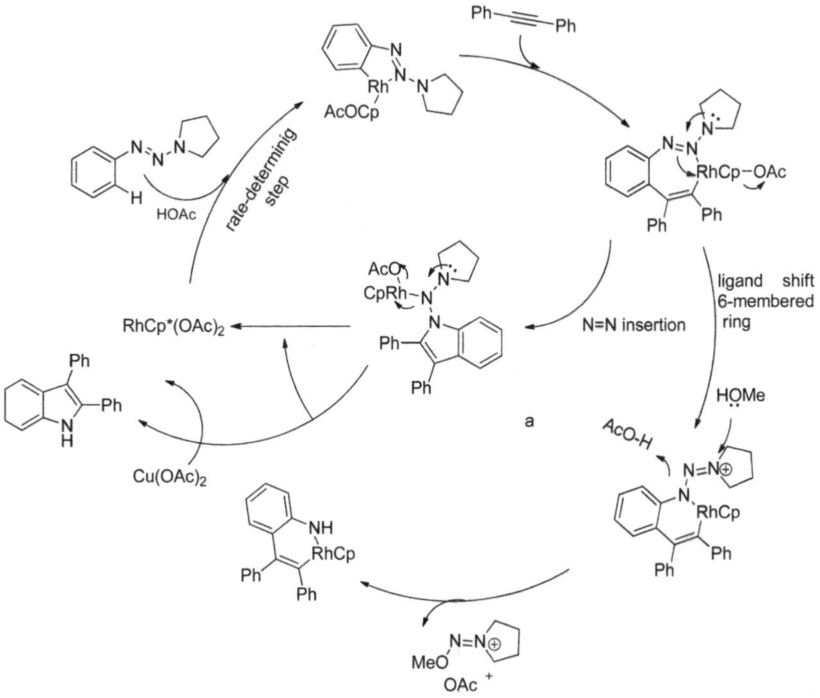

Scheme 4.10 Triazene-directed NH indole synthesis.

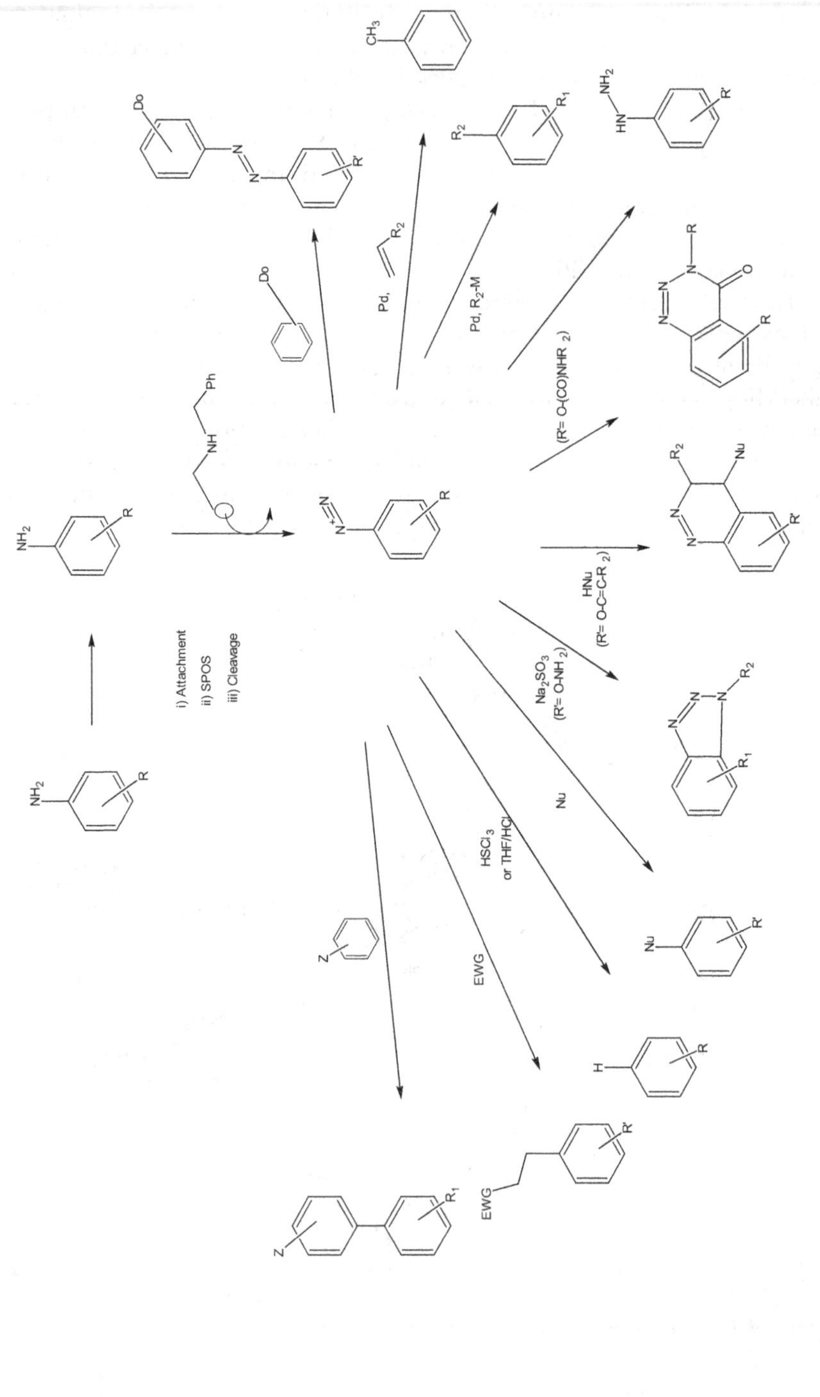

Scheme 4.11 Possibilities of the triazene linker.

linkers will certainly be used for the synthesis of complex organic molecules, such as natural products, on solid supports in the future.

Francesco Marchesi et al. [27] studied the triazenes of clinical interest (i.e., dacarbazine and temozolomide). These triazenes are a group of alkylating agents with similar chemical, physical, antitumor, and mutagenic properties. They also studied their mechanism of action which is mainly related to the methylation of O6-guanine, mediated by methyldiazoniumion, a highly reactive derivative of the two compounds. The cytotoxic/mutagenic effects of these drugs are based on the presence of DNA O6-methylguanine adducts that generate base/base mismatches with cytosine and with thymine. These adducts lead to cell death, or if the cell survives, provoke somatic point mutations represented by C:G→T:A transition in the DNA helix. Triazene compounds have excellent pharmacokinetic properties and limited toxicity. Dacarbazine requires hepatic activation whereas temozolomide is spontaneously converted into an active metabolite in aqueous solution at physiological pH. DTIC activation depends on cytochrome P450-mediated formation of HMTIC, which is subsequently converted into MTIC. On the other hand, TMZ is spontaneously converted into MTIC in aqueous solution at physiologic pH without hepatic activation. Thereafter, the activation pathways derived from DTIC and TMZ are identical. MTIC rapidly tautomerizes into the inactive derivative AIC, that spontaneously produces molecular nitrogen and methyldiazoniumcation, a highly reactive electrophilic ion (Scheme 4.12).

Ana Sousa et al. [28] successfully synthesized a new series of triazene derivatives selective for melanoma cells overexpressing tyrosinase. Metastatic melanoma still remains one of the most difficult cancers to overcome. They exploit the unique enzyme pathway of melanin biosynthesis for the conversion of non-toxic prodrugs into toxic drugs in the melanoma cell. The compounds were designed by coupling two active moieties, the alkylating triazenes and different tyrosinase substrates (Scheme 4.13).

The synthesis and antitumor activity of methyltriazene prodrugs have been studied by Martin J. Wanner et al. [29]. They explored the active resistance of tumor cells against DNA alkylating agents arising by the production of high levels of the DNA repair protein O6-alkylguanine-DNA alkyltransferase (AGT) (Scheme 4.14). This resistance during treatment with, for example, the anticancer agent temozolomide can be reversed by the administration of O6-benzylguanine, a purine that transfers its benzyl group to AGT and irreversibly inactivates it. Stimulated by the favorable therapeutic properties of temozolomide, DNA-methylating triazenes built on the antiresistance benzylguanine ring system were designed and synthesized.

Ningur Noyanalpan et al. [30] studied triazeno derivatives of sulfaguanidine (Scheme 4.15). The sulfaguanidine molecule was chosen by his research group as a carrier and substituted triazeno group was chosen

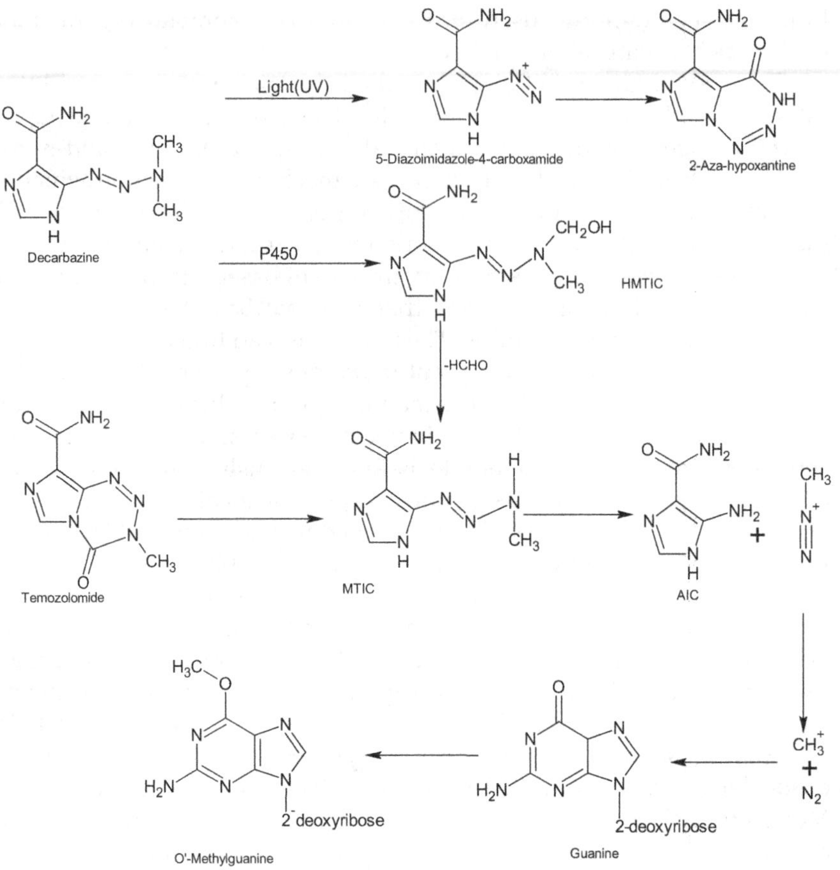

Scheme 4.12 Activation and mechanism of action of DTIC and TMZ.

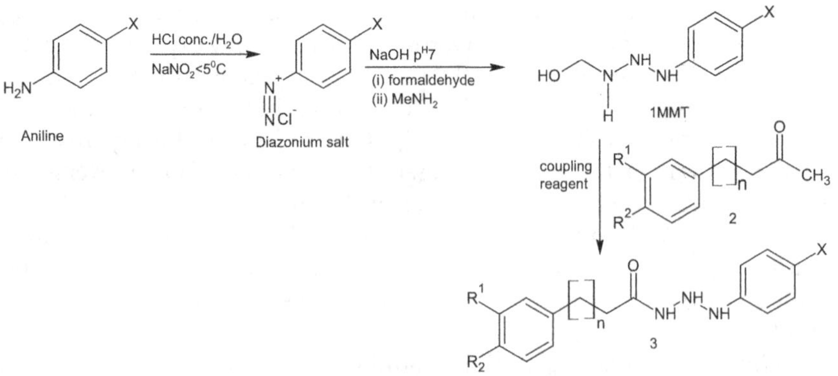

Scheme 4.13 Synthetic pathway of triazene derivatives.

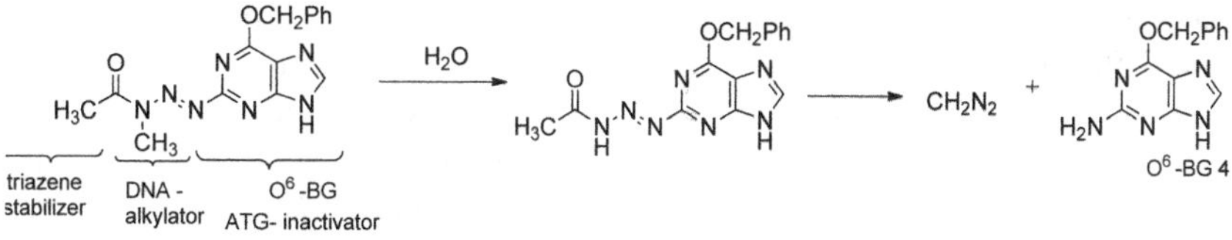

Scheme 4.14 Target triazenes contain an acyl substituent on the triazene to protect the purinyltriazene against premature fragmentation.

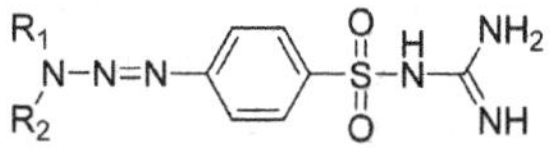

Scheme 4.15 Triazeno derivatives of sulfaguanidine.

as a carcinostatic moiety and they were incorporated. One of the major biochemical properties of neoplastic tissues is their low pH value. This property has been taken into consideration to incorporate a well-known carcinostatic group, substituted triazeno moiety, onto a suitable carrier, sulfaguanidine.

An interesting study has been carried out by James M. Mathews et al. [31] in which 1,3-diphenyl-1-triazene (DPT) is used in the synthesis of polymers and dyes, and has been found as an impurity in the color additives D&C Red 33 and FD&C Yellow 5 (Scheme 4.16). His research group studied the absorption, metabolism, and disposition of 1,3-diphenyl-1-triazene in rats and mice after oral and dermal administration. DPT, randomly labeled in the phenyl rings, was used to investigate its disposition in rodents. Dermal doses to rats and mice (2 and 20 mg/cm²) were poorly absorbed (<7%) in 72 h of exposure. Oral doses of DPT (20 mg/kg) to male rats and mice were well-absorbed and excreted mainly in the urine, with exhalation of volatile organics accounting for about 1% of the dose. The sole volatile component present in breath was benzene, and all

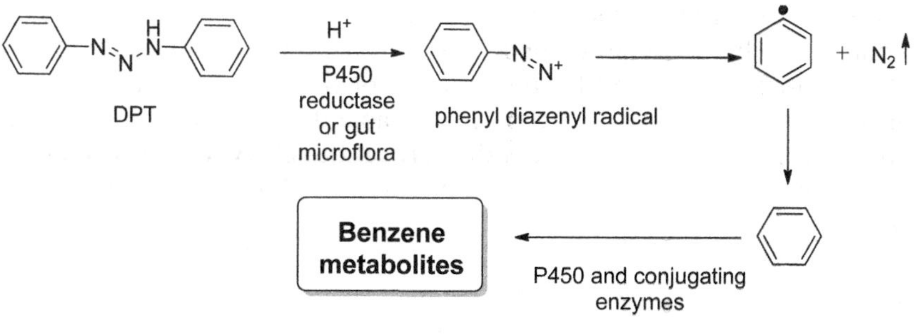

Scheme 4.16 Proposed pathway for the metabolism of DPT.

of the metabolites present in urine were composed of those known for the differential metabolism of benzene and for aniline in the two species. Benzene and aniline were detected in the blood of rats administered with oral doses of DPT, and relatively high circulating levels of their metabolites were measured as early as 15 min post-dosing. Metabolites of these two carcinogens were also formed in human liver slices, indicating a carcinogenic potential for DPT in humans.

Christian Hejesen et al. [32] extended the application of aryltriazenes as traceless linkers for DNA-directed synthesis. After the reaction of one building block with a building block at another DNA strand the triazene linker is cleaved and reduced with hypophosphorous acid in high yield to leave the aryl group with a hydrogen in place of the triazene, that is, without a functional group trace. It was also demonstrated that alternatively the triazene could be converted to an azide, which was used in a cycloaddition reaction. The linker is generally stable at pH >7 and could be stored for several months in a freezer without significant degradation. They have shown that triazene linkers can be applied to link aromatic building blocks to DNA oligonucleotides containing secondary amines. If the aryl groups contain another functional group such as a carboxylic acid it can react with a building block at another DNA strand in a DNA-directed reaction. Subsequently, the triazene linker could efficiently be cleaved and reduced by hypophosphorous acid, leaving the aryl group with a hydrogen in place of the triazene. The overall process is a DNA-directed and traceless transfer of an aromatic building block from one DNA strand to another. They believed that this would find application in the DNA-directed synthesis of encoded chemical libraries. Furthermore, it was shown that the triazene could also be substituted with an azide, and since the Huisgen–Sharpless–Meldal 1,3-dipolar cycloaddition reaction of azides with alkynes is compatible with DNA, the azide approach provides further opportunities for increasing the diversity in such libraries.

For the introduction of the triazene linker directly onto DNA they have synthesized a 40-mer oligonucleotide carrying an internal dT (C6-amino) modifier (O1). Their first approach to the construction of triazene-linked DNA conjugates was to attach a building block already containing the triazene linker to amino-functionalized DNA. However, due to problems related to the presence of two carboxylic acid groups in these derivatives this approach was abandoned. Instead, a strategy in which the triazene is formed directly on the DNA was implemented. First, the amino-modified DNA O1 was linked to the N-Fmoc-piperidine-4-carboxylic acid (Scheme 4.17).

A new series of bis-triazenes has been synthesized by Jeff D. Clarke and co-workers [33] by reacting a series of diazonium salts, bis-secondary amine, and N,N'-dimethyl-1,3-propanediamine (MeNHCH2CH2CH2NHMe), to afford the 3-methyl-3-{3-[1-methyl-3-aryl-2-triazenyl]propyl}-1-aryl-1-

triazenes (Scheme 4.18). A series of related compounds, the 3-ethyl-3-{(*E*)-4-[1-ethyl-3-aryl-2-triazenyl]-2-butenyl}-1-aryl-1-triazenes, have also been synthesized by diazonium coupling with the bis-secondary amine *N,N*′-diethyl-2-butene-1,4-diamine, and the diazonium coupling reaction with *trans-N,N*′-dimethycyclohexane-1,2-diamine has been used to prepare another related series of bistriazenes, the 3- methyl-3-{2-[1-methyl-3-aryl-2-triazenyl]cyclohexyl}-1-aryl-1-triazenes.

Christiane Schotten et al. [34] reported a continuous-flow process for the preparation of triazenes, whereby diazonium salts are generated and converted into their masked or protected triazene derivatives. Key to

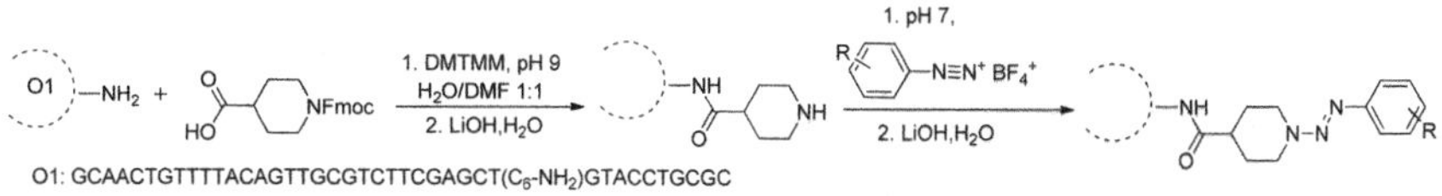
O1
HO
NFmoc
1. DMTMM, pH 9
H2O/DMF 1:1
2. LiOH,H2O
NH
1. pH 7,
N≡N+ BF4+
2. LiOH,H2O
R
O1: GCAACTGTTTTACAGTTGCGTCTTCGAGCT(C6-NH2)GTACCTGCGC

Scheme 4.17 General reaction pathway for coupling an aromatic diazonium salt to a DNA strand *via* a triazene.

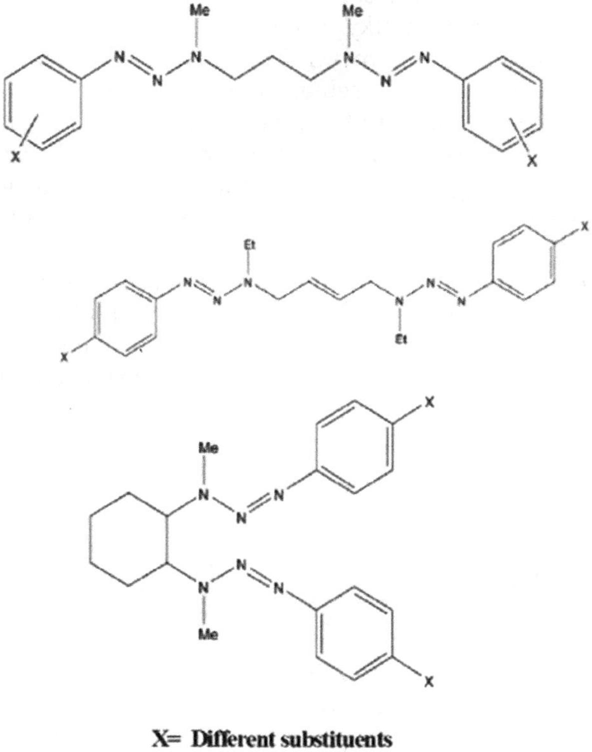
Me
Et
X
X= Different substituents

Scheme 4.18 Synthesis of bis-triazenes.

realizing the process, which is applicable to a wide range of substrates, is the identification of solvent and reagent parameters that avoid fouling and clogging in the tubing used in these studies. The process has also been applied to prepare the antineoplastic agents mitozolomide and dacarbazine. They also reported the isolation and differential scanning calorimetry (DSC) analysis of an anthranilic acid-derived triazene, whose related diazonium salt is a contact explosive. The data highlight improved stability but also suggest that an exothermic process does occur with an onset temperature of 118°C. Finally, an 18-hour continuous operation of the reaction procedure using high-performance liquid chromatography (HPLC) pumps is reported (Scheme 4.19).

4.2   Triazines

Heteracalixaromatics are an emerging generation of macrocyclic host molecules in supramolecular chemistry. As a typical example of heteracalixaromatics, oxacalix[2]arene[2]triazine adopts a shape-persistent 1,3-alternate conformation and can be easily functionalized (Scheme 4.20). Taking it as a platform, a series of oxacalix[2]arene[2] triazine-based amphiphiles bearing long alkyl chains were synthesized by Ruibin Xu et al. [35] through post-macrocyclization functionalization or 3+1 fragment coupling protocols. The self-assembly behavior of these amphiphiles in a mixture of tetrahydrofuran (THF) and water was investigated. Dynamic light scattering (DLS) measurements revealed that the size of the self-assembled aggregates is dependent on the structure of the amphiphiles. The long alkyl chain substituents and/or intermolecular hydrogen bonds were found to promote the self-assembly.

N.C. Desai et al. [36] reported the synthesis of a series of novel 1,3,5-triazine-based thiazole compounds N′-(4-(arylamino)-6-(thiazol-2-ylamino)-1,3,5-triazin-2-yl)isonicotino hydrazides by a series of multistep reactions (Scheme 4.21). Newly synthesized compounds have been characterized

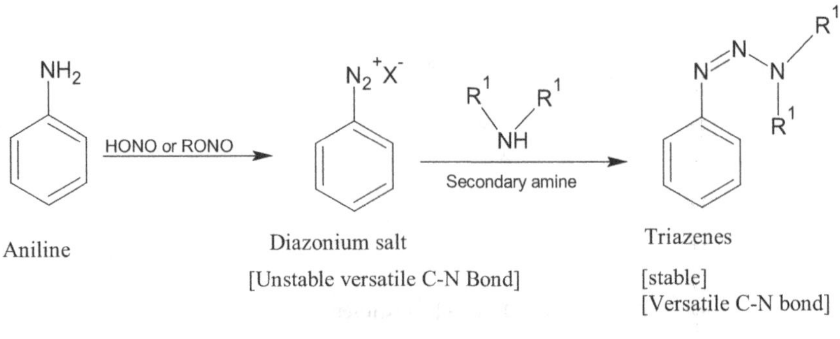

Scheme 4.19 Preparation of triazenes from diazonium compounds.

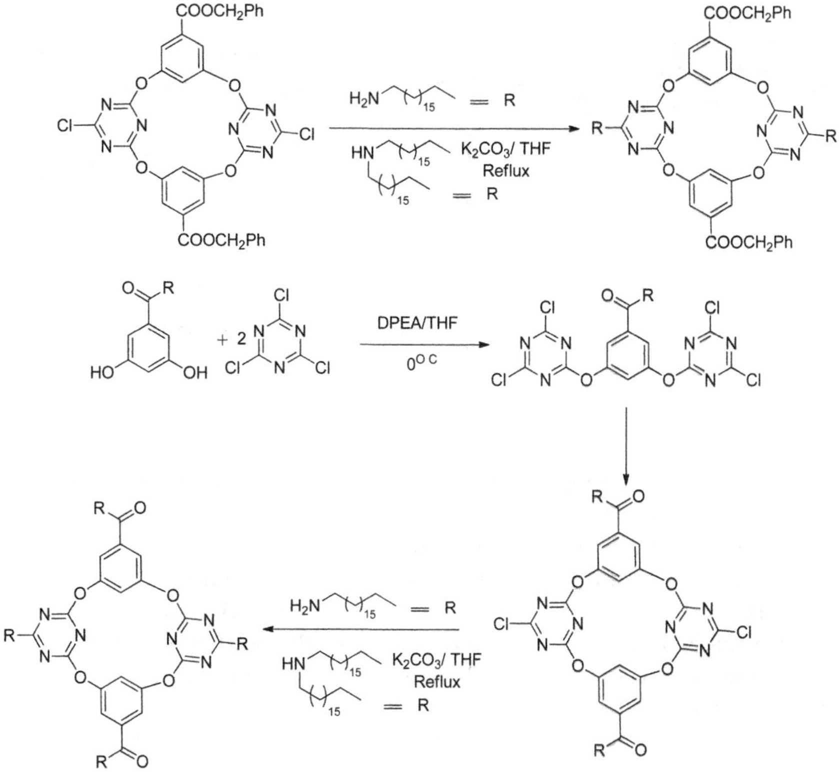

Scheme 4.20 Synthesis of oxacalix[2]arene[2]triazine-based amphiphilic molecules.

by IR, 1H NMR, 13CNMR, and mass spectral data. The antimicrobial screening of title compounds was examined against Gram-positive bacteria (*Staphylococcus aureus, Streptococcus pyogenes*), Gram-negative bacteria (*Escherichia coli, Pseudomonas aeruginosa*), and three fungi (*Candida albicans, Aspergillus niger, Aspergillus clavatus*) by using the serial broth dilution method. Synthesized compounds showed potent inhibitory action against test organisms. Few screened compounds were associated with considerably higher antibacterial and antifungal activities than commercially used antibiotics.

On the basis of structure–activity relationship (SAR) studies, the results suggested that the antimicrobial activity of triazine derivatives was markedly influenced by various substituents on the benzene ring and the incorporation of the electron-withdrawing group caused an enhancement in activity against most test microorganisms. Compounds 3h, 3k, and 3l possessing electron-withdrawing substituents (F and NO_2) on the aromatic ring are more active against all tested panel of bacteria and fungi

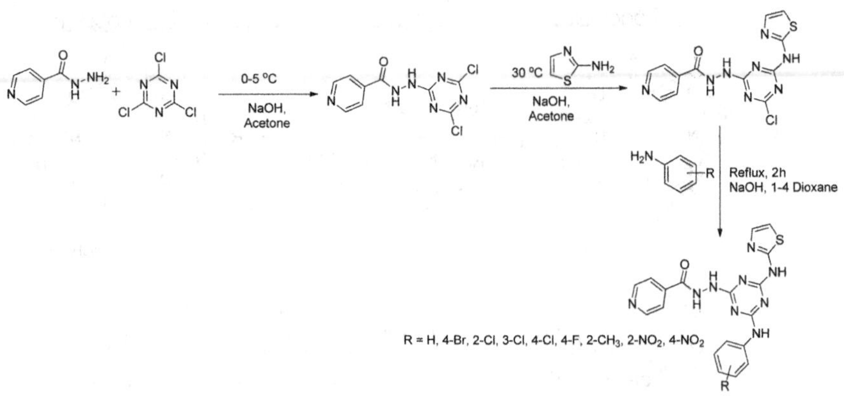

Scheme 4.21 Synthetic route for the 1,3,5-triazine-based thiazole compounds.

than compounds having electron-donating groups. Therefore the electron-withdrawing groups induced a positive effect and the electron-donating groups induced a negative effect on the antimicrobial activity. The preliminary in vitro antibacterial and antifungal screening results of new 1,3,5-triazine-isonicotino hydrazide-based thiazole derivatives exhibited a remarkable antimicrobial potency. The newly synthesized compounds presented hereby significantly differ in their corresponding antimicrobial activity depending on the type of substituents. In the course of this study, particularly derivatives possessing electron-withdrawing groups such as fluoro and nitro were identified as exhibiting potent groups for antimicrobial activity against tested microorganisms. It may be concluded that fluoro and nitro substituent bearing derivatives are the most suitable compounds for achieving the best antimicrobial spectrum. Thus, it may be considered as a promising lead for the further design and development of new chemical entities.

1,3,5-Triazines with tree identical groups: benzotriazol-1/2-yl, imidazol-1-yl, pyrazolyl-1-yl, 3,5-dimethylpyrazolyl-1-yl, 4,5-diphenylimidazol-1-yl, benzimidazolyl-1-yl, 2-methylbenzimidazolyl-1-yl, or 2-phenylbenzimidazolyl-1-yl, were synthesized by Viktor Milata et al. [37]. Their biological activity against wood-destroying fungi *Serpulal acrymans*, *Coniophora puteana*, and *Trametes versicolor* was tested by the impregnated filter paper method. *S. lacrymans* occurred as the most sensitive fungus (of the three fungi) in the presence of triazines. Triazines having three imidazole or three 4,5-diphenylimidazol groups were slightly more effective than others. However, their efficacy in comparison with the commercial fungicides Tebuconazole and IPBC was insufficient (Scheme 4.22).

K. Sztanke et al. [38] reported the synthesis, crystal structure, and antiproliferative activity of novel derivatives of methyl and ethyl

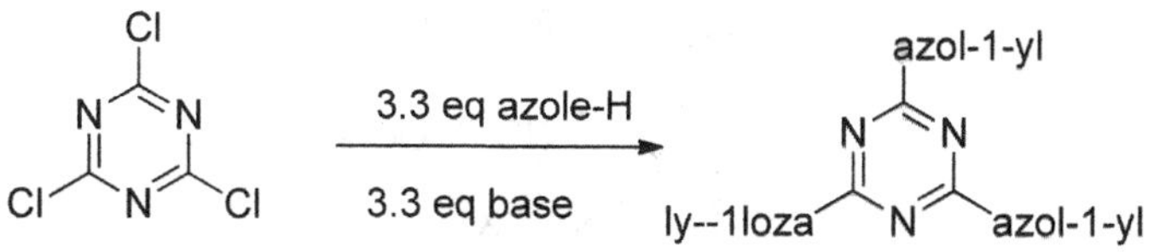

Scheme 4.22 Synthesis of 1,3,5-triazines.

2-(4-oxo-8-aryl-2*H*-3,4,6,7-tetrahydroimidazo[2,1-c][1,2,4]triazin-3-yl)ace-
tates from biologically active 1-aryl-2-hydrazinoimidazolines (Figure 4.3).

The simple and efficient [4+2] domino annulation reactions were
developed for the synthesis of 1,2,4-triazine derivatives by Dong Tang
et al. [39]. The strategies exhibit a high performance with moderate to high
yields by using easily available materials including ketones, aldehydes,
alkynes, secondary alcohols, and alkenes, and represent a powerful tool
for the formation of potentially biological active derivatives. In these
transformations, SeO$_2$ or iodine sources was used as oxidants without
using transition metal catalysts and show good tolerance with moderate
to high yield. This transformation yielded 1,2,4-triazine skeletons, which
can be potentially applied to afford a series of biologically activity derivatives (Scheme 4.23).

Pooja Mullick et al. [40] synthesized a series of 1,2,4-triazine derivatives (Figure 4.4) and evaluated them for their anti-anxiety and

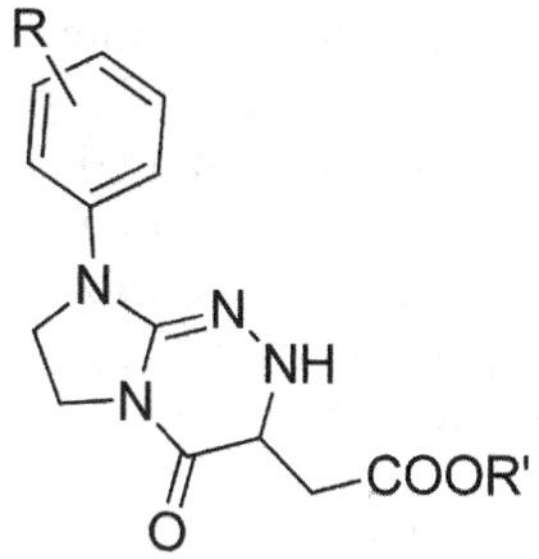

Figure 4.3 Methyl and ethyl 2-(4-oxo-8-aryl-2*H*-3,4,6,7-tetrahydroimidazo[2,1-c]
[1,2,4]triazin-3-yl)acetates.

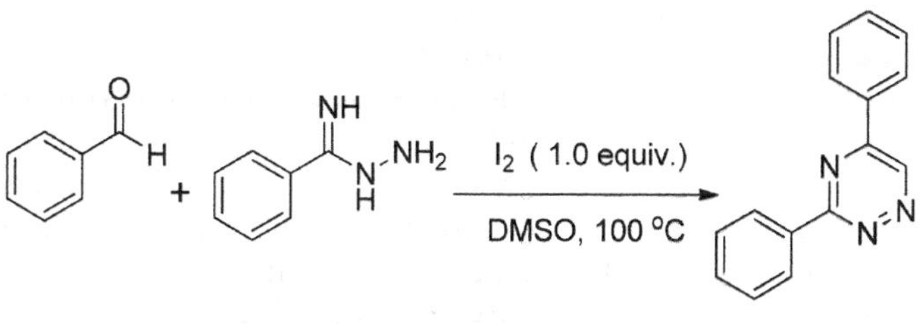

Scheme 4.23 Synthesis of 1,2,4-triazines.

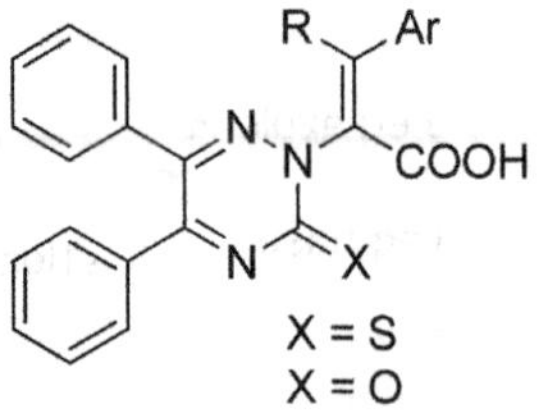

Figure 4.4 1,2,4-triazine derivatives showing anti-anxiety and anti-inflammatory activities.

anti-inflammatory activities. The structures of the synthesized compounds were confirmed on the basis of their spectral data. Many of the triazine compounds were found to possess good activity. In particular, compounds bearing a sulfur atom showed better activity than those bearing an oxygen atom.

All the triazine derivatives obtained were screened for their anti-anxiety and anti-inflammatory activity. The results indicated that for an anti-anxiety activity the dihydroxy substituent on the phenyl ring at the 3, 4-positions yields the compound with better activity than the *para-*, *meta-*, and *ortho*-substituted phenyl rings. Also, the dihalogenated phenyl rings showed an increase in % immobility. Besides, the replacement of the benzene ring with other heterocyclic moieties resulted in better activity. Further, the compounds bearing a sulfur atom showed better activity than the compounds bearing an oxygen atom.

Muhammad Nadeem Arshad et al. [41] report a comparative theoretical and experimental study of four triazine-based hydrazone derivatives. The hydrazones are synthesized by a three-step process from commercially available benzil and thiosemicarbazide. The experimental geometric parameters and spectroscopic properties of the triazine-based hydrazones are compared with those obtained from density functional theory (DFT) studies.

A synthetic procedure has been developed by Z. Pourghobadil et al. [42] for the preparation of a new triazine-derived macrocycle (13,2 9-diphenyl-4,7,20,23-tetraoxa-1,10,12,14,16,17,26,28,30,32-decaaza-[10,10] (2,6)triazinophane). The formation of some transition and heavy metal complexes with the macrocycle was investigated in acetonitrile solution conductometrically at different temperatures. The formation constants of the resulting 1:1 complexes were determined from the molar conductance–mole ratio data and found to vary in the sequence $Hg^{2+} >$ $Pb^{2+} > Cd^{2+} \sim Ag^+ > Cu^{2+} > Tl^+ > Co^{2+} > Ni^{2+} > Zn^{2+}$. The enthalpy and entropy of the complexation reactions were determined from the temperature dependence of the formation constants. In all cases, the complexes were found to be enthalpy-stabilized but entropy-destabilized (Scheme 4.24).

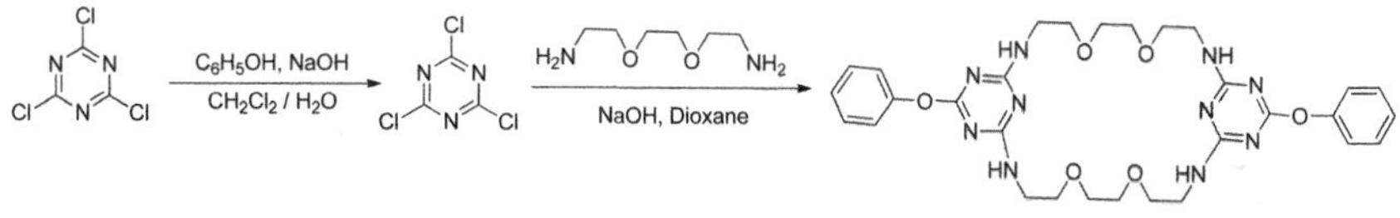

Scheme 4.24 Synthesis of triazine-derived macrocycle.

Pandey et al. [43] developed a three-step triazine synthesis (Scheme 4.25). The reaction of 7-hydroxy-4-methyl coumarin with amido/imido alcohols in ethanol containing concentrated hydrochloric acid afforded 8-aralkyl amido/imido-alkyl-7-hydroxy-4-methyl-coumarins. The interaction of substituted coumarins with hydrazine hydrate in pyridine resulted in 1-amino-8-aralkyl amido/imido-alkyl-7-hydroxy-4-methyl-2-oxo-quinolines. The treatment of these quinolines with formaldehyde in ethanol resulted in 1,3,5-tris-(8-aralkyl amido/imido-alkyl-7-hydroxy-4-methyl-2-oxo-quinolinyl)-2,4,6-hexahydro-s-triazines.

Sym-triazines have been documented for their antitumor, antimicrobial, and anticancer actions. Ashish K Singh and co-workers [44] synthesized a variety of 1,3,5-triaryl-1,3,5-hexahydrotriazine using ultrasound-assisted reactions from the reactions of aryl amines with aqueous formaldehyde (Scheme 4.26). Furthermore, they carried out a special study on the different solvents and revealed that a mixture of water and ethanol was found to be the best solvent.

S-triazine derivatives based quinoline shows a wide range of biological activity. N. Kavitha et al. [45] used 4,7-dichloroquinoline as a starting material and treated it with ethylene diamine, which afforded 4-substituted 7-chloroquinoline. This was further reacted with 1,5-disubstituted

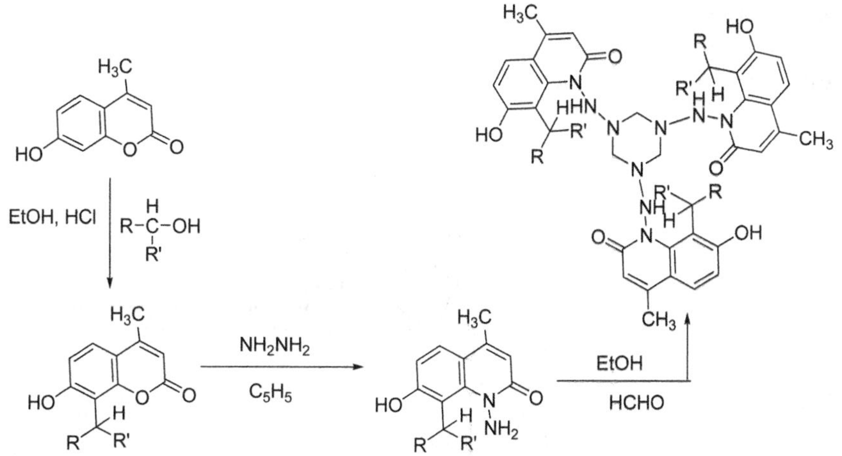

Scheme 4.25 Three-step triazine synthesis.

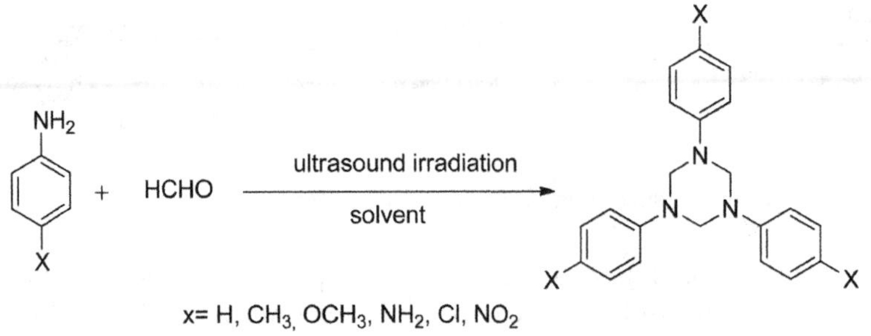

x= H, CH₃, OCH₃, NH₂, Cl, NO₂

Scheme 4.26 Synthesis of Sym-triazines.

cyanuric chloride yielding 1,3,5-triazine chloroquinoline derivatives. All the synthesized compounds were characterized using IR, 1H,13C NMR, mass spectral studies, and elemental analysis. The final compounds were screened for their antibacterial activity using *E. coli, S. aureus*, and *S. typhi* and antifungal activity (Figure 4.5).

Mohamed Abd El-LatifZein et al. [46] synthesized a series of triazine derivatives to assess their anti-proliferation efficacy on human cancer cell lines. So, 2-(amino) thioxo-3-phenyl-1,2,5,6-tetrahydro-1,2,4-triazine-6-one was prepared *via* the reaction of N-benzoyl glycine with thiosemicarbazide under fusion at 130°C. The acetylation and alkylation of 2-(amino) thioxo-3-phenyl-1,2,5,6-tetrahydro-1,2,4-triazine-6-one with acetic anhydride and ethyl chloroacetate yielded the corresponding N-acetamide derivative and ethyl N-aminoacetate derivative, respectively (Figure 4.6). The cytotoxic activities of the 1,2,4-triazine-6-one derivatives were studied on the tumor cell lines, human colon carcinoma (HCT-116), and human hepatocellular carcinoma cells (HepG-2) using the MIT viability test. The result showed that the investigated compound, 1-(p-chlorophenyl)-4-thioxo-5-phenyl-triazino[2,1-a]-7,8-dihydro-1,2,4-triazine-8-one, had a significantly great cytotoxic effect compared to that of the other compounds.

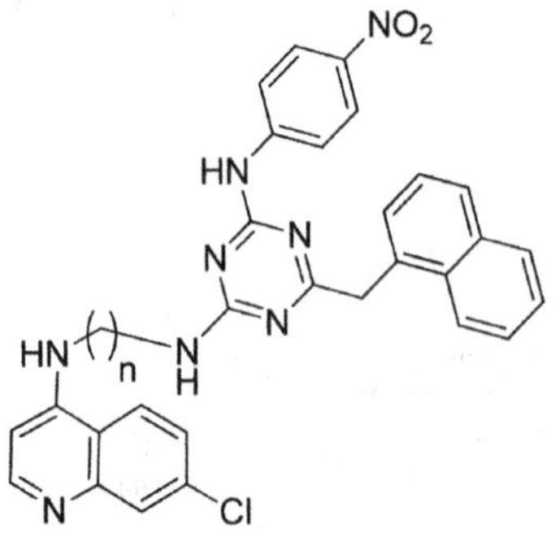

Figure 4.5 1,3,5-triazine chloroquinoline derivative.

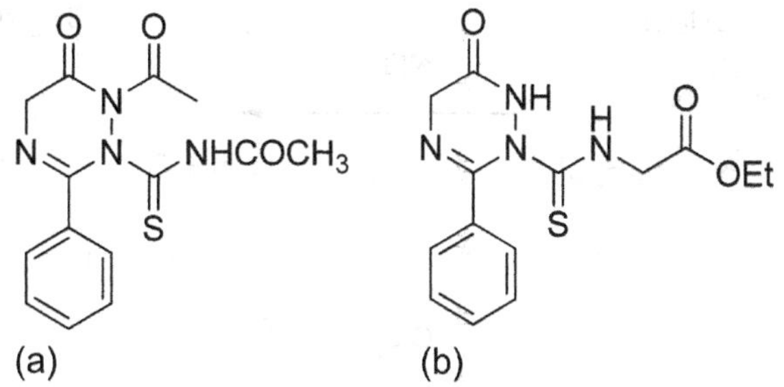

Figure 4.6 (a) N-acetamide derivative and (b) ethyl N-aminoacetate derivative.

The design, synthesis, and biological evaluation of a series of 6-aryl-1,6-dihydro-1,3,5-triazine-2,4-diamines are described by Anna C.U. Lourens et al. [47]. These compounds exhibited in vitro antiplasmodial activity in the low nanomolar range against both drug-sensitive and drug-resistant strains of *P. falciparum*, with 1-(3-(2,4-dichlorophenoxy)propyl)-6-phenyl-1, 6-dihydro-1,3,5-triazine-2,4-diamine hydrochloride identified as the most potent compound from this series against the drug-resistant FCR-3 strain (IC50 2.66 nM). The compounds were not toxic to mammalian cells at therapeutic concentrations and were shown to be inhibitors of parasitic DHFR in a biochemical enzyme assay (Scheme 4.27).

They designed and synthesized a series of flexible analogues of cycloguanil bearing an aromatic substituent at the 6-position of the dihydrotriazine ring. The compounds displayed potent activity against both drug-resistant and drug-sensitive forms of *P. falciparum* in a whole-cell in vitro assay, and were shown to act as inhibitors of parasitic DHFR in a biochemical enzyme assay. Good activities against the mutant enzyme (Ala16Val+Ser108Thr DHFR) were due to the flexibility of the 5-side chain which can avert steric hindrance introduced by the Ser108Thr mutation,

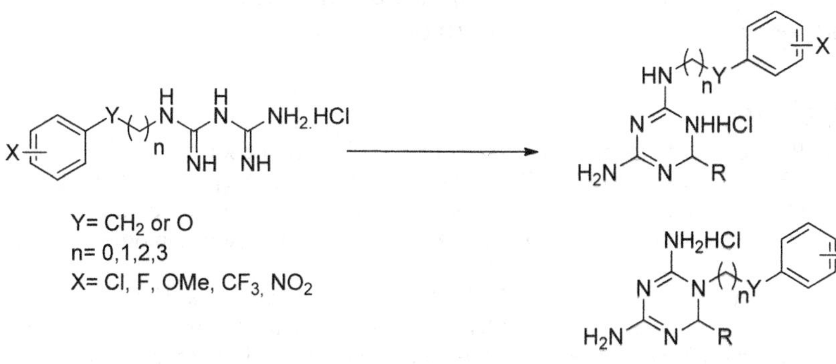

Y= CH$_2$ or O
n= 0,1,2,3
X= Cl, F, OMe, CF$_3$, NO$_2$

Scheme 4.27 Synthesis of 6-aryl-1,6-dihydro-1,3,5-triazine-2,4-diamines.

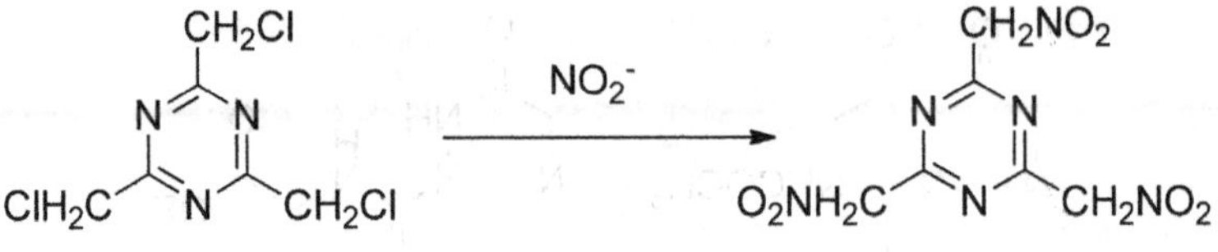

Scheme 4.28 Synthesis of nitromethyl derivatives of 1,3,5-triazines.

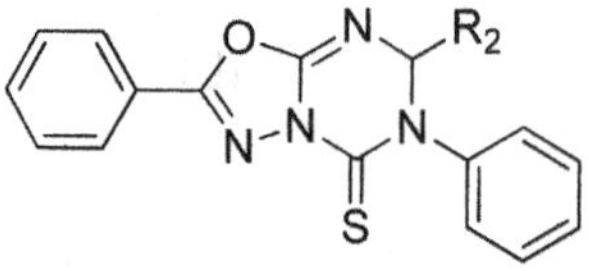

Figure 4.7 Substituted aryl-1,3,4-oxadiazolo-[3,2-a]-1,3,5-triazine derivative.

while the 6-aryl substituent, in contrast to the 6,6-dimethyl groups of cycloguanil, can avert steric hindrance caused by the Ala16Val mutation. This is indicative of the potential application of these compounds as antimalarial leads for further development.

V. Shastin et al. [48] studied the synthesis of nitromethyl derivatives of 1,3,5-triazines (Scheme 4.28). The reaction takes place *via* the destructive nitration of 2,4,6-tris[di(carboxy)methylene]hexahydro-1,3,5-triazine and its esters. A first representative of the nitromethyl derivatives of 1,3,5-triazine, viz. 2,4,6-tris(nitromethyl)1,3,5-triazine, has been synthesized.

Basedia et al. [49] studied bioactive heterocyclic rings in which 1,3,4-oxadiazole and 1,3,5-triazine are fused to give substituted aryl-1,3,4-oxadiazolo-[3,2-a]-1,3,5-triazine derivatives. The synthesized compounds were evaluated for antimicrobial activity against a variety of bacterial and fungal strains, and some of these compounds have shown significant antibacterial and antifungal activity (Figure 4.7). In their research, a series of novel fused heterocyclic compounds 1,3,4-oxadiazolo-[3,2-a]-1,3,5-triazine derivatives has been successfully synthesized by bridging between the 1,3,5-triazine nucleus, which is one of the active leads present in many standard drugs.

References

1. Rouzer, C. A., Sabourin, M., Skinner, T. L., Thompson, E. J., Wood, T. O., Chmurny, G. N., Klose, J. R., Roman, J. M., Smith, R. H. Jr., and Michejda, C. J. Oxidative metabolism of 1-(2-chloroethyl)-3-alkyl-3-(methylcarbamoyl)triazenes: Formation of chloroacetaldehyde and relevance to biological activity. *Chem. Res. Toxicol.* 1996, 9, 172.
2. (a) Connors, T. A., Goddard, P. M., Merai, K., Ross, W. C. J., and Wilman, D. E. V. Tumor inhibitory triazenes: Structural requirements for an active metabolite. *Biochem. Pharmacol.* 1976, 25, 241–246; (b) Hickman, J. A. Investigation of the mechanism of action of antitumourdimethyltriazenes. *Biochimie* 1978, 60, 997.

3. Nicolaou, K. C., Boddy, C. N. C., Li, H., Koumbis, A. E., Hughes, R., Natarajan, S., Jain, N. F., Ramanjulu, J. M., Brase, S., and Solomon, M. E. Total synthesis of vancomycin—Part 2: Retrosynthetic analysis, synthesis of amino acid building blocks and strategy evaluations. *Chem. Eur. J.* 1999, 5, 2602.
4. Brase, S., Dahmen, S., and Pfefferkorn, M. Solid-phase synthesis of urea and amide libraries using the T2 triazene linker. *J. Comb. Chem.* 2000, 2, 710.
5. Jones II, L., Schumm, J. S., and Tour, J. M. Rapid solution and solid phase syntheses of oligo(1,4-phenylene ethynylene)s with thioester termini: Molecular scale wires with alligator clips. Derivation of iterative reaction efficiencies on a polymer support. *J. Org. Chem.* 1997, 62, 1388.
6. Moore, J. S. Shape-persistent molecular architectures of nanoscale dimension and references therein. *Acc. Chem. Res.* 1997, 30, 402.
7. Wirshun, W., Winkler, M., Lutz, K., and Jochims, J. C. 1,3-Diaza-2-azoniaallene salts: Cycloadditions to alkynes, carbodiimides and cyanamides. *J. Chem. Soc. Perkin Trans.* 1998, 2, 1755.
8. (a) Jason Krutz, L., Shaner, D. L., Weaver, M. A., Webb, R. M., Zablotowicz, R. M., Reddy, K. N., Huang, Y., and Thomson, S. J. Agronomic and environmental implications of enhanced s-triazine degradation. *Pest Manage. Sci.* 2010, 66, 461; (b) Li, D., Zhang, Z., Li, N., Wang, K., Zang, S., Jiang, J., Yu, A., Zhang, H., and Li, X. Dispersant-assisted dynamic microwave extraction of triazine herbicides from rice. *Anal. Methods* 2016, 8, 3788.
9. Kazuya, K., Nobuhiro, K., Kohtaro, T., Akira, T., Aiko, O., Hitoshi, K., Ko, W., Peter, Böger. Novel 1,3,5-triazine derivatives with herbicidal activity. *Pest Manage. Sci.* 1999, 55 (6), 642.
10. (a) Shah, D. R., Modh, R. P., and Chikhalia, K. H. Privileged s-triazines: structure and pharmacological applications. *Future Med. Chem.* 2014, 6, 463–477; (b) Singla, P., Luxami, V., and Paul, K. Triazine as a promising scaffold for its versatile biological behavior. *Eur. J. Med. Chem.* 2015, 102, 39.
11. (a) Safin, D. A., Holmberg, R. J., Burgess, K. M. N., Robeyns, K., Bryce, D. L., and Murugesu, M. Hybrid material constructed from Hg(NCS)2 and 2,4,6-tris(2-pyrimidyl)-1,3,5-triazine (TPymT): Coordination of TPymT in a 2,2'-bipyridine-like mode. *Eur. J. Inorg. Chem.* 2015, 441; (b) Manzano, B. R., Jalón, F. A., Soriano, M. L., Carrión, M. C., Carranza, M. P., Mereiter, K., Rodríguez, A. M., de la Hoz, A., and Sánchez-Migallón, A. Anion-dependent self-assembly of silver(i) and diaminotriazines to coordination polymers: non-covalent bonds and role interchange between silver and hydrogen bonds. *Inorg. Chem.* 2008, 47, 8957; (c) Vicente, A. I., Caio, J. M., Sardinha, J., Moiteiro, C., Delgado, R., and Félix, V. Evaluation of the binding ability of tetraaza[2]arene[2]triazine receptors anchoring L-alanine units for aromatic carboxylate anions. *Tetrahedron* 2012, 68, 670; (d) Zhu, X., Mahurin, S. M., An, S. H., Do-Thanh, C. L., Tian, C., Li, Y., Gill, L. W., Hagaman, E. W., Bian, Z., Zhou, J. H., Hu, J., Liu, H., and Dai, S. Efficient CO_2 capture by a task-specific porous organic polymer bifunctionalized with carbazole and triazine groups. *Chem. Commun.* 2014, 50, 7933; (e) Santos, M. M., Marques, I., Carvalho, S., Moiteiro, C., and Felix, V. Recognition of Bio-bio-relevant dicarboxylate anions by an azacalix[2]arene[2]triazine derivative decorated with urea moieties. *Org. Biomol. Chem.* 2015, 13, 3070; (f) Wang, Z. N., Wang, X., Yue Wei, S., Xiao Wang, J., Ying Bai, F., Heng Xing, Y., and Xian Sun. Triazine–polycarboxylic acid complexes: synthesis, structure and photocatalytic activity. *New J. Chem.* 2015, 39, 4168.

12. (a) Manzano, B. R., Jalón, F. A., Soriano, M. L., Rodríguez, A. M., de la Hoz, A., and Sánchez-Migallón, A. Multiple hydrogen bonds in the self-assembly of aminotriazine and glutarimide. Decisive role of the triazine substituents. *Cryst. Growth Des.* 2008, 8, 1585; (b) Yagai, S. Supramolecularly engineered functional π-assemblies based on complementary hydrogen-bonding interactions. *Bull. Chem. Soc. Jpn.* 2015, 88, 28.
13. Al-Saadawy, N. H., and Faraj, H. R. Synthesis and characterization of a new triazene complexes for Cu(II), Ni(II), Co(II), Zn(II) and Fe(II). *Am. Chem. Sci. J.* 2016, 10, 1.
14. Garcia, C. T., Pulido, D., Albericio, F., Royo, M., and Nicolas, E. Triazene as a powerful tool for solid-phase derivatization of phenylalanine containing peptides: Zygosporamide analogues as a proof of concept. *J. Org. Chem.* 2014, 79, 11409.
15. Lohse, J., Schindl, A., Danda, N., Williams, C. P., Kramer, K., Kuster, B., Witte, M. D., and Medard, G. Target and identify: Triazene linker helps identify azidation sites of labelled proteins via click and cleave strategy. *Royal Soc. Chem.* 2017, 53(87), 11929.
16. Schotten, C., Aldmairi, A. H., Sagatov, Y., Shepherd, M., and Browne, D. L. Protected diazonium salts: A continuous-flow preparation of triazenes including the anticancer compounds dacarbazine and mitozolomide. *J. Flow Chem.* 2016, 6, 218.
17. Unsala, S., Cikla, P., Kucukguzel, S. G., Rollas, S., Sahin, F., and Bayrak, O. F. Synthesis and characterization of triazenes derived from sulphonamides. *Marmara Pharm. J.* 2011, 15, 11.
18. Stebani, J., and Nuyken, O. Synthesis and characterization of a novel photosensitivetriazene polymer. *Makromol. Chem. Rapid Commun.* 1993, 14, 365.
19. Gross, M. L., Blank, D. H., and Welch, W. M. The triazene moiety as a protecting group for aromatic amines. *J. Org. Chem.* 1993, 58, 2104.
20. Pawelec, W., Aubert, M., Pfaendner, R., Hoppe, H., and Wilen, C. E. Triazene compounds as a novel and effective class of flame retardants for polypropylene. *Polym. Degrad. Stab.* 2012, 97, 948.
21. Calvalho, E., Iley, J., Perry, M. J., and Rosa, E. Triazene drug metabolites: Synthesis and plasma hydrolysis of anticancer triazene containing amino acid carriers. *Pharm. Res.* 1998, 15, 931.
22. Carvalho, E., Francisco, A. P., Iley, J., and Rosa, E. Triazene drug metabolites: Synthesis and plasma hydrolysis of acyloxymethyl carbamate derivatives of antitumour triazenes. *Bioorg. Med. Chem.* 2000, 8, 1719.
23. Little, V. R., Tingley, R., and Vaughan, K. Triazene derivatives of (1,*x*)-diazacycloalkanes. Part Xa. synthesis and characterization of a series of 1,4-Di[2-aryl-1-diazenyl]-*trans*-2,5-dimethylpiperazines. *Can. J. Chem.* 2014, 1.
24. Nuyken, O., Stebania, J., Lippert, T., Wokaun, A., and Stasko, A. Photolysis, thermolysis, and protolytic decomposition of atriazene polymer in solution. *Macromol. Chem. Phys.* 1995, 196, 751.
25. Wang, C., Sun, H., Fang, Y., and Huang, Y. General and efficient synthesis of indoles through triazene-directed C–H annulation. *Angew. Chem. Int. Ed.* 2013, 52, 5795.
26. Brase, S. The virtue of the multifunctional triazene linkers in the efficient solid-phase synthesis of heterocycle libraries. *Acc. Chem. Res.* 2004, 37, 805.

27. Marchesi, F., Turriziani, M., Tortorelli, G., Avvisati, G., Torino, F., and Vecchis, L. D. Triazene compounds: Mechanism of action and related DNA repair systems. *Pharmacol. Res.* 2007, 56, 275.
28. Sousa, A., Santos, F., Gaspar, M. M., Calado, S., Pereira, J. D., Mendes, E., Francisco, A. P., and Perry, M. J. The selective cytotoxicity of new triazene compounds to human melanoma cells. *Bioorg. Med. Chem.* 2017, 25(15), 3900.
29. Wanner, M. J., Koch, M., and Koomen, J. N. Synthesis and antitumor activity of methyltriazene prodrugs simultaneously releasing DNA-methylating agents and the antiresistance drug O6-benzylguanine. *J. Med. Chem.* 2004, 47, 6875.
30. Noyanalpan, N., Ozden, S., and Ozden, T. Triazene derivatives: Sulfaguanidine as a carrier for substituted triazeno group. *J. Fac. Pharm Ankara.* 1977, 7, 104.
31. Mathews, J. M., and Costa, K. S. D. Absorption, metabolism and disposition of 1,3-diphenyl-1-triazene in rats and mice after oral and dermal administration. *Am. Soc. Pharmacol. Exp. Therap.* 1999, 27, 1499.
32. Hejesen, C., Petersen, L. K., Hansenb, N. J. V., and Gothelf, K. V. A traceless aryl-triazene linker for DNA-directed chemistry. *Org. Biomol. Chem.,* 2013, 11, 2493.
33. Clarke, J. D., Moser, S. L., and Vaughan, K. Synthesis and characterization of a series of 3-methyl-3-{3-[1-methyl-3-aryl-2-triazenyl]propyl}-1-aryl-1-triaz enes and related compounds. *Can. J. Chem.* 2006, 84, 831.
34. Schotten, C., Aldmairi, A. H., Sagatov, Y., Shepherd, M., and Browne, D. L. Protected diazonium salts: A continuous-flow preparation of triazenes including the anticancer compounds dacarbazine and mitozolomide. *J. Flow Chem.* 2016, 6, 218.
35. Xu, R., Hou, B., Wang, D., and Wang, M. Synthesis and self-assembly of novel oxacalix[2]arene[2]triazineamphiphiles. *Sci. China Chem.* 2016, 59, 1306.
36. Desai, N. C., Makwana, A. H., and Rajpara, M. K. Synthesis and study of 1,3,5-triazine basedthiazole derivatives as antimicrobial agents. *J. Saudi Chem. Soc.* 2016, 20, 334.
37. Milata, V., Reinprecht, L., and Kizlink, J. Synthesis and antifungal efficacy of 1,3,5-triazines. *Acta Chimica Slovaca* 2012, 5, 95.
38. Sztanke, K., Pasternak, K., Rzymowska, J., Sztanke, M., and Szerszen, M. K. Synthesis, structure elucidation and identification of antitumoural properties of novel fused 1,2,4-triazine aryl derivatives. *Eur. J. Med. Chem.* 2008, 43, 1085.
39. Tang, D., Wang, J., Wu, P., Guo, X., Li, J. H., Yang, S., and Chen, B. H. Synthesis of 1,2,4-triazine derivatives via [4+2] domino annulations reactions in one-pot. *RSC Advances* 2016, 6(15), 12514.
40. Mullick, P., Khan, S. A., Begum, T., Verma, S., Kaushik, D., and Alam, O. Synthesis of 1,2,4-triazene derivatives as potential anti-anxiety and anti-inflammatory agents. *Acta Poloniae Pharmaceutica.* 2009, 66, 379.
41. Arshad, M. N., Bibi, A., Mahmood, T., Asiri, A. M., and Ayub, K. Synthesis of triazine-based hydrazone derivatives. *Molecules* 2015, 20, 5851.
42. Pourghobadil, Z., Majidi, F. S., Asli, M. D., Parsa, F., Moghimi, A., Ganjalil, M. R., Aghabozorgand, H., and Shamsipur, M. Synthesis of a new triazine derived macrocycle and a thermodynamic study of its complexes with some transition and heavy metal ions in acetonitrile solution. *Polish J. Chem.* 2000, 74, 837.

43. Pandey, V. K., Tusi, S., Tusi, Z., Joshi, M., and Bajpal, S. Synthesis and biological activity of substituted 2,4,6-s-triazines. *Acta Pharm.* 2004, 54, 1.
44. Singh, A. K., Shukla, S. K., and Quraishi, M. A. Ultrasound mediated green synthesis of hexa-hydro triazines. *J. Mater. Environ. Sci.* 2011, 2, 403.
45. Kavitha, N., Karthi, A., Arun, A., and Shafi, S. S. Synthesis, characterisation and antimicrobial activity of some novel s-triazene derivatives incorporating quinolone moiety. *Acta Chim. Pharm. Indica.* 2016, 6, 53.
46. El-LatifZein, M. A., and El-Shenawy, A. I. An efficient synthesis of 1,2,4-triazine-6-one derivatives and their *in vitro* anticancer activity. *Int. J. Pharmacol.* 2016, 12, 188.
47. Lourens, A. C. U., Gravestock, D., Zyl, R. L. V., Hoppe, H. C., Kolesnikova, N., Taweechai, S., Yuthavong, Y., Kamchonwongpaisan, S., and Rousseau, A. L. Design, synthesis and biological evaluation of 6-aryl-1,6-dihydro-1,3,5-triazine-2,4-diamines as antiplasmodialantifolates. *Org. Biomol. Chem.* 2016, 14, 7899.
48. Shastin, V., Godovikova, T. I., Golova, S. P., Khmel'nitskii, L. I., and Korsunskii, B. L. Nitromethyl derivatives of 1,3,-triazene: Synthesis and properties. *Chem. Heterocycl. Comp.* 1997, 33, 1095.
49. Basedia, D. K., Shrivastava, B., Dubey, B. K., and Sharma, P. Synthesis, characterization and antimicrobial activity of novel substituted aryl- 1,3,4-oxadiazolo-[3,2-a]-1,3,5-triazine derivatives. *Int. J. Drug Deliv.* 2013, 5, 379.

Index

A

A 549 cell line, 84
Acetic acid induced writhing test, 42, 43
Acyclic triazenes, 81
3-Acyloxymethyloxycarbonyl-1-aryl-3-methyltriazenes, 87
Adsorption, distribution, metabolism, and excretion (ADME), 21, 36
AGT, *see* Alkyltransferase
Ala16Val mutation, 106
Alkyltransferase (AGT), 93
3-Aminoacyl-1-aryl-3-methyltriazenes, 87
Analytical reagents, hydroxytriazenes, 3, 4
Antimicrobial screening, 99–100
6-Aryl-1,6-dihydro-1,3,5-triazine-2,4-diamines, 105
1-Aryl-3,3-dimemethyltriazenes, metabolism pathways, 88
1-Aryl-3-methyltriazenes, 87
Aryl-1,3,4-oxadiazolo-[3,2-a]-1,3,5-triazine derivative, 106
Azide-bearing molecules, 83

B

Biomolecular targets, 5
BirA, 83, 84
Bis-triazene synthesis, 96, 97
Buchheim, Rudolf, 17

C

CADD, *see* Computer-aided drug design
Click and cleave strategy, 83
Clinker resin synthesis, 83, 84
Computer-aided drug design (CADD), 17, 40
Continuous-flow protocol, 84, 97

Cupferron, 1
Cyclic triazine, 81
Cyclodepsipeptide derivatives, 83

D

Dacarbazine, ix, 93
DCC method, 87
Density functional theory (DFT), 102
DHFR, 105
Diazonium compounds, triazenes, 84, 98
Differential scanning calorimetry (DSC), 86, 98
Diminazene aceturate, 18
1,4-Di-[2-aryl-1-diazenyl]-*trans*-2,5-dimethylpiperazines, 87
structure, 87
1,3-Diphenyl-3-hydroxytriazene, PASS of, 22–36
1,3-Diphenyl-1-triazene (DPT), 95, 96
metabolism pathway, 95
DLS, *see* Dynamic light scattering
DNA binding, 18, 45
DPT, *see* 1,3-Diphenyl-1-triazene
Drosophila melanogaster, 43, 44
Drug discovery process, 19, 21, 81
Drug likeliness, 5
DSC, *see* Differential scanning calorimetry
DTIC, 93
activation and action mechanism of, 94
Dynamic light scattering (DLS), 98

E

Electron-donating group, 100
Electron-withdrawing group, 99–100
Ethyl N-aminoacetate derivative, 104, 105

111